Blue Ridge 2020

Steve Nash

BLUE RIDGE 2020
An Owner's Manual

The University of
North Carolina Press
Chapel Hill and London

Designed by Heidi Perov
Set in Berkeley Book and Tekton
by Keystone Typesetting, Inc.

Manufactured in the United States of America

The paper in this book meets the guidelines for
permanence and durability of the Committee on
Production Guidelines for Book Longevity of the
Council on Library Resources.

Library of Congress Cataloging-in-Publication Data

Nash, Steve, 1947–
    Blue Ridge 2020: an owner's manual / Steve Nash.
        p.   cm.
    Includes bibliographical references (p.   ) and index.
    ISBN 0-8078-4759-3 (pbk.: alk. paper)
    1. Ecology—Blue Ridge Mountains. 2. Ecosystem
management—Blue Ridge Mountains. I. Title.
QH104.5.B5N37   1999
577′.09755—dc21   98-20106
                                   CIP

03 02 01 00 99   5 4 3 2 1

For my wife and hiking partner, Linda

If you would prepare for the future,
prepare to be surprised!
—Kenneth Boulding, presidential address to
   the American Economics Association, 1984

We can expect to be surprised by the future.
But we don't have to be utterly dumbfounded.
—Kenneth Boulding, in a seminar
   at Lakewood, Colorado, ca. 1978

CONTENTS

Solutions

Figures

Plates

Color plates appear following page 12.

For time-consuming help with every sort of question I am deeply indebted to the scores of scientists and other experts who provided information for this book. One hundred thirty-five of them are named in the bibliography, including more than two dozen who contributed lengthy interviews for the Solutions sections of the book, but many others also gave abundant help, without hesitation. I owe them all fervent thanks. In an era of harassed schedules, pinched budgets, and seeming public hostility—and even during the weeks when the federal government was "out of business" in 1995—the public servants of the U.S. Forest Service, the National Park Service, and the Environmental Protection Agency were consistently generous.

A great many of those who helped me with research results and interviews also reviewed excerpts for accuracy. Peter S. White of the University of North Carolina spent many hours reviewing a draft, as did James C. Spaulding, a mentor and a fine science writer. Jack C. Schultz of Pennsylvania State University carefully read, and took strong exception to parts of, the initial chapters. Rick Webb, Tom Rawinski, Lynnell Reese, Will Orr, Bill Wallner, Robert Zahner, Stuart Pimm, Michael Pelton, Katherine J. Elliott, and Philip Bell endured hours of questions on the road, on the phone, and on the trail. Arthur Chappelka, Niki Nicholas, James Renfro, Joe Mitchell, Paul Delcourt, William Wallner, Byron Freeman, Noel Burkhead, Cindy Huber, Art Bulger, Julie Thomas, Christopher Lucash, Bill McShea, John Rappole, Kerry Rabenold, Scott Robinson, Sam Droege, Bruce Peterjohn, Dave Plunkett, Sharon Mohney, Steven Oak, David Wear, Jim Loesel, Jim Sisler, Peter Weigl, Samuel McLaughlin, Michael Pelton, Tom Blount, and Shepard Zedaker commented helpfully on lengthy excerpts.

The support of UNC Press editor David Perry is keenly appreciated. He lent encouragement from this book's earliest beginnings and kept the faith. Editors Pam Upton and Grace Buonocore helped keep a sprawling manuscript aligned and incorporated countless refinements. I am grateful in addition to the University of Richmond, where I teach, for the priceless gift of time, as well as a telephone and a quiet place to work—the fundamentals needed to commit an act of journalism. University of Richmond librarians Nancy Vick, Bill Sudduth, Melanie Hillner, and Betty Tobias were cheerful and efficient in processing a sizable pile of requests for scarce journals, lost books, and crumbling government documents over more than two years. University statistician Kevin Beam provided ample help with making sense of bird and human population data.

The maps that appear on these pages were derived from data and images gathered by the U.S. Forest Service for its ecosystem mapping project and by the Southern Appalachian Assessment, a state and federal multiagency effort completed in 1996. Combining the two sources of images would not have been possible without the help of Chris Frye, of the U.S. Forest Service. The data supplied in the text of the Southern Appalachian Assessment was also indispensable, and I have cited it many times. John Molenar of Air Resource Specialists put meticulous effort into the visibility images that appear in this book, on the strength of nothing more than my request. Permission to use the illustration in figure 11 was granted by Randall Arendt of the Natural Lands Trust. Permission to use poll data in Solutions 21, from the Institute for Research in Social Science data archive at the University of North Carolina, Chapel Hill, was granted by Beverly Wiggins, associate director. Permission to use material from "Beyond Sprawl" was granted by Julie M. Hovan, administrative manager at the Bank of America's Environmental Policies and Programs Group.

In expressing thanks to John Harmon and Carrie Teegardin of the *Atlanta Journal and Constitution*, Rex Springston of the *Richmond Times-Dispatch*, Stan DeLozier of the *Knoxville News-Sentinel*, and E. A. Torriero of the *Fort Lauderdale Sun-Sentinel* for the chance to piggyback on some of their fine reporting, I need to explain why their work is not cited directly: I contacted and interviewed the sources whose names appeared in their stories myself for additional information and to confirm the accuracy of the handful of quotations appearing in this book that have been published earlier. The close reading, suggestions, and moral support of my friends and colleagues Hank Nuwer and Mike Spear were crucial, as were conversations with my uncles Lowell Nash and Lester Briggs and my cousin Howard Nash. They helped dampen the urge to preach to the choir.

Affectionate thanks to my wife, Linda, for her tenacious and skillful editing and for seeing this project through. And last and most of all, thanks to her and to my children, Alex, Forrest, and Celeste, for all the times we've spent together in the mountains.

The many people whose assistance I've relied on do not, of course, necessarily agree with the views expressed in this book. Errors discovered here are, despite all this deeply appreciated help, my own.

This book is about what may happen to the natural systems of the mountains we call the Blue Ridge during the coming decades. While writing it, I've guessed three things about you. You're wary of, but not cynical about, pronouncements on the environment; you care about the mountains for any of several possible reasons; you don't have much time.

Few of us do, and we all prefer that issues be reduced and promptly registered on that time-saving "bottom line." Yet we're disappointed when terse, seemingly authoritative summaries turn out to be too simple, too general, too easily called into question by still other flat statements. These pages may try your patience a bit but not, I hope, unintentionally. The broad outlines of the future seem clearly lit at some points, hidden in a fog of uncertainty at others. In embracing that uncertainty I try to give a fuller, though perhaps less entertaining, account.

As to your skepticism: I have chosen to ignore much valuable and interesting information available from both environmentalists and industry sources. I have relied instead almost exclusively on the published work of, and interviews with, scientists and other professionals in academia or in government. Where disagreement exists within the sources of scientific information, I have tried to present it fairly.

This attempt to glean "neutral" information may not satisfy readers whose opinions are already well settled. Scientific research can, of course, be controversial, and criticism of government experts is a commonplace from all points on the political spectrum. They are "job-killers" if you don't like the EPA, and coat-holders for the timber industry if you don't like the U.S. Forest Service. So no source is above criticism, but the academic and government research and opinion I have relied on at least has the virtue of relative independence of outside financing and control.

It may seem presumptuous, if not preposterous, to say what could occur in the Blue Ridge even in the near future. But I'll cast my lot with the late economist and seer Kenneth Boulding, quoted above. There is no way around the need to look ahead. Whenever we and our government make or avoid decisions that affect the natural world, we are gauging what is likely to happen in the decades to come. Careful research can help inform those decisions by indicating a range of likely possibilities. It does not pretend to lay out prophecy.

Careful research is also wrong on a regular basis, especially when it tries to cope with the future. Imagine each of the studies and commentaries that appear in these pages as part of an instrument panel, and that you're a pilot in the cockpit of an airliner. It's basically a room full of meters, telling how much fuel you have, which direction you're headed, how fast you're mov-

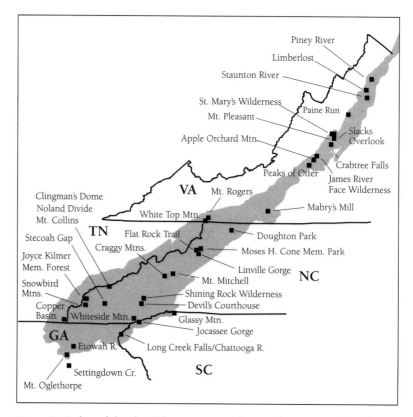

Figure 1. *Outline of the Blue Ridge ecosystem, with some places mentioned in the text (locations approximate). (Sources:* Keys et al., *Ecological Units of the Eastern United States,* CD-ROM; Hermann, *Southern Appalachian Assessment GIS Data Base,* CD-ROM)

ing, what's up ahead, your altitude, and a hundred other things. Suppose you know, too, that an unknown number of these dials and gauges are giving false readings. Do you make your way out to the main cabin, to watch the movie and wait for whatever happens next? Or do you stay to piece the information together as best you can and act on it?

It is no disservice to their authors to say that some of the research studies included here will, with time, turn out to be busted meters. We are mistaken to expect anything else of science, since science is always tentative.

Readers must also reckon with the illusion of precision that is generated by numbers. Properly regarded, the poll data, or nesting success among thrushes, or concentrations of sulfate pollution noted in this book are not descriptions of reality. They carry nothing like the certainty of numbers describing, say, the fingers on your left hand or the fans in a baseball stadium. Instead, they are general indications—smudges of data on a spectrum of probability. And finally, know that information of this kind has an

Figure 2. *Some towns and cities in the Blue Ridge that are mentioned in the text (locations are approximate).* (*Sources:* Keys et al., *Ecological Units of the Eastern United States*, CD-ROM; Hermann, *Southern Appalachian Assessment GIS Data Base*, CD-ROM)

unknown shelf life. The passage of time may not diminish the odds that what you read is truth, but it rarely improves them.

This doesn't mean that the numbers and the research are worthless but that their promise is somewhat different: taken together, they can help us get our bearings. As anyone who has ever embarked on a long and somewhat uncertain journey knows, a sense of the right direction can be crucial. On what better basis can we make decisions about the future?

I have presented what I think is most important and most nearly true, but this account is objective neither in the sense of neutrality nor of equal time for all views. It includes some strong opinions about how best to manage the future of the Blue Ridge. The sources of these views defy pigeonholing, and their ideas often overlap to a surprising degree. My own outlook should be plainly stated: the natural systems of the mountains are of critical importance for us, and in their own right. We can afford to sustain them; we can't afford not to.

Whether we choose to be pilots or passengers is, in our political system, an individual reckoning. We can't suppose, though, that such matters will somehow be outside our control. Individual citizens, acting together, will continue to have huge impacts on the Blue Ridge—its forests and wildlife, its rivers and skies, and its future—every day.

Blue Ridge 2020

When I saw them yesterday from two miles overhead, these mountains were becalmed, changeless. They filled the horizon so completely that only a bolt of sun, glancing off a hidden creek far below, signified the plane's motion.

This morning down along Snowbird Creek is also a lengthening pause. A light mid-April snow has made spring hold its breath. The birds are silent, the hardwoods dead gray. Icicles suspend from the rocks along the trail. The landscape seems outside time, enduring, and safe from interruption.

The illusion of changelessness, fed by the scale of the forest and, perhaps, our own fond wish, is one reason why we seek out these mountains so avidly. The Snowbird range is part of the Blue Ridge, a big natural backyard for the East and Midwest, visited by at least thirty million of us each year and inhabited by nearly three million more. But despite the stillness of the hour, this region is undergoing profound changes now.

Some are the expression of natural forces that have always been. The icicles will melt by noon, and a lime-green tide of new leaves rises higher along the ridges each day. On a slower timetable, the mountain rivers ease more deeply into their beds and shift their curves some with every new flood. Populations of cedar waxwings, earthworms, and bobcats swell or decline. The forest around them ages or, ripped open by fires and storms, regrows. In a thousand years, Blue Ridge granites and greenstones weather down patiently by half an inch, while the whole continent they are part of slides westward 20 yards.

But other changes, more frequent now, register human influence on the land. Some are unhidden. At the junction of Snowbird and Sassafras Creeks, the wreck of what looks like a '39 Dodge panel truck rusts away peaceably under some hemlocks. And along these trails, just two or three hand spans measure the girth of the tree trunks, because they are only about sixty years old. Underfoot, the half-buried ties of a dismantled logging railroad, from which iron spikes still jut here and there, help explain why the forest is so young.

A few miles farther west and north, high up on Burntrock Ridge, a forgotten grove of never-cut "old-growth" hemlock and poplar shows what the measure of the forest must once have been. Each of these trees would take several human arm lengths, and life spans, to compass.

As the surrounding square miles of young forest and old roads in this part of North Carolina attest, we humans are numerous, and busy. One geomorphologist estimates that we rearrange more billions of tons of surface earth and rock each year with bulldozers, dynamite, and erosion than any single natural force—glaciers, rivers, shoreline currents, or the collisions of continental plates.

From all this power, powerful choices arise. In the Blue Ridge, we have chosen over the years to raze the woods, exterminate elk, wolf, deer, and buffalo, poison rivers with mine and mill waste, graze and plow much of the deforested land to ruin. But more recently, we've allowed the return of millions of acres of trees, clarified some of the rivers, reestablished deer and bear and at least a few falcons, river otters, and red wolves. We are making choices of the same magnitude today, with even deeper consequences in the next century. And so the question arises: how will these mountains be changed then—in, say, the year 2020 and after?

One kind of answer comes from scientists who study natural systems. Their work gives us a blurred, fragmentary, and contingent—but useful— mosaic of the future. Maps can help us piece it together. Various kinds of boundaries define the Blue Ridge, and the maps showing towns, roads, and county lines are the most familiar. But a map conjured only recently out of field studies and computer databases has become increasingly important for how we think about, and plan for, the natural landscape. In it, the boundaries of the Blue Ridge are based on ecosystems: the way communities of plants and animals relate to the land, the climate, and each other.

This reorganizes our outlook in an unaccustomed way. We usually focus on environmental issues state by state, agency by agency, species by species, chemical by chemical. The ecosystem map draws our attention instead to a complex, interdependent natural region. Then the role of each pollutant or agency or species is more distinctly, and logically, seen: each of them is a factor affecting the well-being of the whole Blue Ridge.

Smaller ecosystems nest within larger ones like Chinese boxes. At the largest view, the whole planet is an ecosystem—a set of interdependent natural processes. So is the tiny community of rare filmy ferns, mosses, pennyworts, saxifrage, mock orange, and dwarf dandelions kept moist by the spray among the rocks of South Carolina's remote Long Creek Falls. And the whole Blue Ridge can also be thought of as an ecosystem, because of its distinct tapestry of elevations, climate types, rock strata, and vegetation. They differentiate this long chain of mountains sharply from the flatlands to the east and the ridges and valleys to the west.

On the ecosystem map the 17,000-square-mile Blue Ridge region extends from northern Georgia to South Mountain, Pennsylvania. It is shaped like a ragged, 550-mile-long fish, 80 miles wide at the head and only 10 miles across at the narrow point of the tail. It is the core of the larger Southern Appalachian region (plates 1, 2, 3).

Folded within the two dozen or so mountain ranges that make up the Blue Ridge are a myriad of natural habitats: shale barrens and river gravels; caves, bogs, and balds; oak-hickory, beech gap, and Table Mountain pine forests among them. They are home to 29 kinds of snakes, 70 or more species of mammals, a couple of hundred kinds of birds, more than 1,400 different flowering plants, at least 70 species of fish, and more than 130 tree species—nearly as many as in all of Europe.

Geography and geohistory are the sources of much of this biological richness. The 2-mile-thick glaciers of the most recent ice age never reached south of Pennsylvania. So, affected but not obliterated by the climate, plant life to the south is more various and more intricately related. Because of its length, the Blue Ridge intersects the ranges of both southerly and northerly plant and animal communities. Its rise from a few hundred feet above sea level to frigid heights—fifty-two peaks are 6,000 feet or higher—has roughly the same effect, nurturing a complex and various mix of warm- and cold-adapted species.

Even the forest floor, sometimes as vivid as a coral reef, startles. You might see an immense bracket fungus that looks like a burnished copper platter as it emerges from a dead hickory. Clumps of mushrooms fluoresce in hues of salmon and yellow. Furtive skinks and salamanders show loud blues and scarlets. "If you go anywhere along the Blue Ridge you find strikingly colored fishes," says icthyologist Robert Jenkins. "The Appalachians have been a center of evolution of aquatic fauna. The diversity is incredible, and incredibly beautiful."

And those are just the readily visible organisms. All this profusion of life—the range of ecosystems and species and the genetic variations within species—is known as biological diversity, or biodiversity. It is fostered by time and slow change, as species compete, adapt, evolve new forms, or die out.

Plans for its future have even higher significance because we citizens paid for, own, and control so much of this natural treasure through our government. Within the Blue Ridge ecosystem are seven national forests, two national parks linked by the Blue Ridge Parkway and the Appalachian Trail, twenty-nine officially designated wilderness areas, and many more state parks, forests, and preserves—the largest concentration of public lands east of the Mississippi.

The future health of the Blue Ridge ecosystem is worth money:

- Visitors to the two national parks alone spend roughly $350 million locally each year, creating more than twelve thousand jobs. The "ripple effect" of this income stream makes its impact on local economies much greater.
- A single year's visitors spent about $475 million in the communities along the Blue Ridge Parkway in the mid-1990s, and that rate was 30

percent higher than in the decade before. The money plus its "ripple effects" yielded nearly $300 million in local income—about thirteen thousand jobs paying an average of $22,500 per year. Respondents to a survey said that the chance to "observe the beauty of nature" was the most important benefit of their parkway trip.

- Wood products, mostly from privately owned lands, generated an average annual payroll of $800 million and forty-six thousand jobs in the Blue Ridge each year from 1978 to 1994.

But we are learning to think of the world in terms of "ecosystems" and "biodiversity" partly because so many living things, and the relationships that support them, are disappearing. Some species are now extinct that lived, only a century or two ago, in the larger region that the Blue Ridge is part of. They include a range of animals from wood bison to four species of birds—the ivory-billed woodpecker, Bachman's warbler, the passenger pigeon, the Carolina parakeet—and many aquatic species and insects.

The Blue Ridge itself shelters 13 animal and 18 plant species that appear on the federal "endangered" and "threatened" lists. Thirty-six other animals and 122 plants in the Blue Ridge are considered to be species whose survival is "of concern." For aquatic species, no separate Blue Ridge tally is available, but about four dozen varieties of fish, mollusks, water insects, amphibians, and invertebrates in the larger Southern Appalachian region are also listed as "endangered" and "threatened."*

Extinctions have always been part of natural history, since life first arrived on the planet. The fossil record shows that there have been a handful of "cataclysmic" extinction episodes such as the era, sixty-five million years ago, when the dinosaurs disappeared. Between these events, however, species die out at a steadier natural or "background" rate.

Here's a roughly estimated arithmetic for this "background" extinction between cataclysmic events: on average, a species will persist for about a million years. So the chance of losing a particular species in any given

*Government biologists who administer those lists say they lag well behind reality. At the direction of Congress, consideration of listings of endangered species was suspended for more than a year during 1995–96 (see U.S. Fish and Wildlife Service, " 'Candidates' for Endangered Species List"). A "monitor list" of species that might be threatened or endangered—but had not been adequately researched and should be watched—was abolished by the Fish and Wildlife Service, which administers the Endangered Species Act, in 1996 (see U.S. Fish and Wildlife Service, "Revised List of 'Candidates,'" 1–4). Officials said then that the agency was "severely short of money and overwhelmed by a backlog of hundreds of imperiled species" (see Cushman, "Moratorium on Protecting Species Is Ended," A1). Monitoring in the Blue Ridge is so spotty and populations fluctuate so widely, U.S. Fish and Wildlife Service rare species biologist Nora Murdock says, that other Blue Ridge species could be declining seriously without our knowledge.

century is only about one in ten thousand—making it a "very improbable event," University of Tennessee theoretical ecologist Stuart Pimm says.

Against that background rate of extinction, we can estimate in broad terms how forcefully humankind is affecting biodiversity. About four hundred species are endemic to the Blue Ridge ecosystem—that is, they are found nowhere else. It is our great good fortune that, at least in recent times, we may not have lost any Blue Ridge endemics. But a third of them—ninety-one plants and forty-two animals—are now listed by the Nature Conservancy as "critically imperiled," "imperiled," or "vulnerable to extinction."

We can't predict how many on this long list of "at risk" species will survive the coming decades, because we can't predict how strenuous our own efforts to protect them will be. But if only one becomes extinct at our hands by the year 2100, Pimm calculates, the human-caused rate of extinction will be 25 times greater than the average natural, or background, rate. Or, if you prefer, 2,500 percent greater. If thirteen Blue Ridge endemics do not survive—about 10 percent of the "at risk" group—that rate of extinction would be 325 times faster than the natural rate, or 32,500 percent faster.

Pimm describes some of his own work as that of an ecological ambulance chaser, trying to salvage "terminally beaten-up" species. He recently returned to his home on the Knoxville side of the Smokies from Kenya. The comparison was stark.

"In most of the other parts of the world they've lost their forests, or they're losing them," he says. "We in the East may have the finest deciduous forests left on the planet. In the future, this may be one of the only parts of the world where you can walk in forests for days at a time.

"But we don't want to gloss over our history. . . . The traces of our past are really obvious. As for species diversity, we have no way of knowing how much we've lost. We'll probably never know."

Extinction is an incremental process. In the way we humans sense the passage of time, it proceeds slowly at first. By the time many species are declared "endangered," federal officials say, the chances for their long-term survival are nearly foreclosed because more than 90 percent of their habitat has already been destroyed.

Noel Burkhead of the National Biological Service* studies biodiversity

---

*The research staff of the National Biological Service was created from deep cuts in the science staff of the National Park Service, the Bureau of Land Management, and the U.S. Fish and Wildlife Service in 1995. In 1996 the NBS itself was abolished as a separate entity by Congress and made a part of the U.S. Geological Survey. Its budget was cut 20 percent for the following year (see Baker, "How Science Fared," 10). A bill was introduced in the same session of Congress to abolish the U.S. Geological Survey.

among freshwater fish along the Blue Ridge. "We're nickel-and-diming habitat at all levels of the ecosystem," he says. "The direction all this is leading is that ultimately, we're going to be losing more and more species. I consider that to be a sad prophecy, but I don't think I'm going out on a limb. We're talking about major fractions of entire classes of animals.

"I fully believe that a hundred years from now biologists will look at my work and think: 'Wow, he got to hold that species of fish in his hand. That animal no longer exists.'"

If there's an impasse on a federal construction project that may drive a species to extinction, a cabinet-level committee makes the final go/no-go decision. They are the so-called God squad. But in a democracy in which dozens of species are now threatened, we have all become minor deities, with an immediate, if fractional, role in the prospects for their survival.

It's an odd new problem for us. My Uncle Lester's backyard, for example, is overrun with shiny, wriggling earwigs, which make a satisfying crackle under the heels of my two delighted sons. "What good is an earwig?" he wants to know. "What difference would it make if they were extinct?"

The question, applied to one species or another, has probably occurred to most of us. In a section of Virginia's Jefferson National Forest, thick with valuable Blue Ridge timber that the Forest Service had slated for logging, a small population of the Peaks of Otter salamander makes its only earthly home. Four federal agencies have worked out a plan to ensure its survival, and some of the logging will be prohibited. Why such trouble over a salamander?

What we are really asking is, "What good is a (name the rare species) to us humans, right now?" The answers, in sum, are these: individual species can be strikingly useful to us, and useful, too, in holding the whole ecosystem together. Their loss cannot be undone. They have, as a part of Creation or of Evolution, innate value outside our reckoning.

But if we try to meet the what-good-is-it question on its own, narrow terms, the truest answer is this: we can't often justify the continued existence of any single species with a direct, immediate use. We will never be wise enough to know what we will discover next about ourselves, our needs, and how we relate to the life around us.

Naturalist Aldo Leopold wrote famously that if nature, "in the course of aeons, has built something we like but do not understand, then who but a fool would discard seemingly useless parts? To keep every cog and wheel is the first precaution of intelligent tinkering."

There is no mysticism in this. As nearly any scientist will testify, we often have only rudimentary scientific knowledge of the Blue Ridge and other ecosystems despite generations of research. Hints of the inescapable quality of our ignorance abound.

Specialists keep turning up "new" millipedes, beetles, fishes, and flowering plants in the Blue Ridge that have been there for thousands of years at least. Great Smoky Mountains National Park has been studied far more intensively than most natural areas because of its protected status. Yet when specialists arrive to study spiders, insects, or aquatic life, they still sometimes discover previously unknown species. Not just unknown there, but anywhere.

A small, worn-out bungalow in the park affords desk space for researchers in various disciplines. On a bulletin board, they record and update what our species understands so far about the rest of life in the mountains. The list tells only of organisms discovered and named and does not imply that we know much else about them. It is a humbling summary (table 1).

Even bacteria and fungi can turn out to be useful in ways that few could have guessed. The potential pharmaceutical properties of these organisms are largely unexplored. Two thousand varieties of fungi have been identified in Great Smoky Mountains National Park—only a tenth of the estimated total. A single gram of soil in a forest may hold as many as ten thousand species of bacteria.

College students recently found a previously unknown form of the fungus that produces the billion-dollar drug cyclosporin on a beetle grub in woods near their campus. Taxol, part of a new chemotherapy for ovarian and other cancers, was discovered in a type of western yew tree formerly considered a "trash species."

Genetic diversity within species, too, can have immediate uses. Because of an imported insect, for example, the Blue Ridge is on the verge of losing the Fraser fir, which grows nowhere else in the wild. These trees are commercially valuable, grown by the hundreds of thousands on Christmas tree farms. But the firs on Mount Rogers in southern Virginia have shown what may be a form of limited genetic resistance to the insect invasion, and their seeds have attracted growers like a gold strike.

"It's very difficult to keep growers from poaching seed and seedlings on Mount Rogers," says Shepard Zedaker, a forest ecologist at Virginia Tech. "The Forest Service has a heck of a time keeping Fraser fir poachers out. It's a huge problem. They've had a cooperative agreement from time to time with the Christmas tree growers association to try to police it. It's been a real zoo."

Even insects—a little deficient in charm for most of us—also have essential roles in the ecosystem. Entomologist E. O. Wilson calls them "the little things that run the world." Earwigs, for example, are primarily detritivores.

Table 1. Status of Inventories of the Principal Natural Resource Groups—Great Smoky Mountains National Park

Group	Documented No. of Taxa	Estimated No. of Taxa	Mapped % of Taxa
Fungi	2,250	20,000	0
Lichens	393	600–700	50
Bryophytes	467	490–550	1
Vascular plants	1,597 (1,275 native)	1,700–1,800	8
Lepidopterans	724	1,300	0
Ephemeroptera, plecoptera, tricoptera, odonata	365	750	54
Dipterans	Several hundred	2,000–3,000	0
Coleopterans	Couple of hundred	1,000–4,000	0
All other insect orders	Several hundred	5,000–8,000	1
Spiders	376	600–800	1
Mollusks	About 100	200	2
All other invertebrates	A few hundred	Up to several thousand	0
Fish	58	72	
Amphibians	37	40	Less than 15
Reptiles	36	36–38	Less than 5
Birds	240	245	Less than 1
Mammals	67	Fewer than 70	10
Geology	25	25–28	75
Soils	?	?	Less than 1
Natural communities	12	12–25	

Status of Knowledge	Remarks
Fair to poor	Most microfungi unknown but play an important role in forest ecology
Fair to good	
Fair	
Fair to good	Several new taxa discovered each year, more taxa than any other National Parks unit; 130 considered rare, 315 exotic
Most, poor	
Fair to poor	One of the richest areas of its size in North America
Unknown	
Unknown	Incredibly rich group—a few endemics are known
Unknown	
Poor to unknown	Southern Appalachians are a regional center of biodiversity
Poor	
Unknown	
Well known	
Fairly well known	For its size, only areas of Central America rival no. of taxa
Well known	
Well known	Many neotropical species have declined 50 percent in the last 20–30 years, park offers one of the largest refuges from cowbird parasitism; female cowbirds lay up to 40 eggs a season and penetrate up to 4 kilometers into forest
Fairly well known	
80 percent	
Unknown	
Fairly well known	

They consume bits of decaying organic stuff and help to break it down into chemically simpler matter that is good for soil and more easily used by plants.

Another and more colorful detritivore is the Laurel Creek millipede, which probably evolved into its present form between two million and three million years ago. It is black with two rows of lemon-yellow spots along its back and sixty lemon-yellow legs. Found in a patch of maple, oak, hemlock, and dogwood in the neighborhood of Mabry's Mill along the Blue Ridge Parkway, it lives, so far as is known, nowhere else. Its rarity is arguably not the result of human activity. Rather, its kind has been isolated and diminished by time, circumstance, and the long, slow turning of global climate. The hemlocks and dogwoods are under immediate threat from oncoming waves of imported pests, however, and that could easily affect the millipede's chances for survival.

Any such rare species can act, if we are paying attention, as a bell-wether—a warning of possible jolts to the health of the ecosystem. If they are in trouble, the trouble is worth investigating. The late Mollie Beattie, a wildlife biologist and director of the U.S. Fish and Wildlife Service, once sketched this connection:

"The reality is that endangered species are warning flags and smoke alarms of ecological stress. . . . If we are willing to look, we will see them often pointing to . . . incipient economic upheaval. The truth is that our economy depends on the sustained health of our environment. What is economic in the long run is what conserves endangered species. No accurate cost-benefit analysis would calculate in favor of extinction."

Species with a larger ecosystem role than a remnant group of millipedes may also be at risk when ecosystems are broadly disrupted. Even "every-day" plants or insects can become "seldom seen" or "nevermore." The honeybee turned up largely missing from the Blue Ridge and the rest of the eastern United States during the late 1990s. Itself a non-native, the honeybee is now the target of two introduced mites that suck its blood. Honey is the least of it. The bee's loss to agriculture has been called potentially devastating, because it is a pollinator of commercial crops. The effect of its loss as a pollinator in the wild is as yet unknown. We comprehend so little about such relationships that we can seldom predict the consequences of changing them.

The Blue Ridge may be able to do without earwigs or honeybees. Other detritivores and pollinators might take their place. But when population declines or extinctions snip away one part of the web, other strands weaken. This process of ecosystem impoverishment, which biologists call "depauperization" or "pauperization," is fundamental. It reverses the creative process of evolution, driving life backward into diminished complexity, stability, and adaptability.

Life goes on in the waters of Settingdown Creek, near the north Georgia foothills, for example, but it's a different kind of life. This pauperized stream is polluted by silt eroded from poorly managed construction sites nearby. "There's activity in silted-up streams," Noel Burkhead says, "but what you have is a complex system reduced to a few tolerant organisms.

"Species are the nuts and bolts of all ecology. They contribute to and are part of the processes that drive the whole ecological engine. If the rate of extinction is being accelerated, you're losing the nuts and bolts. Obviously, at some point, you're going to start having major breakdowns."

Similar warnings about the importance of biodiversity have sometimes been dismissed as speculative or overwrought—a merely fashionable insistence that every bird and bug is critically important.

But whenever the interdependent nature of nature is pooh-poohed, some bit of startling news turns up. A team of ecologists recently compared test plantings—miniature ecosystems—that contained more and less biodiversity. They measured growth, resistance to drought and disease, and the soil's ability to prevent nutrients from washing away.

By all three yardsticks, the plots with more species stayed healthier. More biodiversity generates better odds for survival of the ecosystem as a whole. And the reverse is also true, the scientists reported: the results demonstrate that the loss of species threatens ecosystem functioning and long-term survival.

Ethicist Mark Sagoff argues, though, that we value places like the Blue Ridge for reasons other than their utility. "It's like old people and children. Old people aren't going to earn any income, although you love them. Children only had economic value in the old days when child labor was permitted, and you could actually get children to earn income.

"There's always been this intense difficulty about finding any economic value for objects of moral worth, or love, or cultural and aesthetic concern. Because the difference—between the merely commercial or useful and the loved, the appreciated, the object of moral responsibility—is the oldest distinction in the world."

Biodiversity has been characterized as a vast library—eons of life histories, evolutionary experience recapitulated in the genetic code of each species. No part of it is replaceable, once lost. Having begun to decipher a few volumes in the Blue Ridge and elsewhere, we risk losing others for all time: their beauty, their evolutionary legacy, their innate worth independent of our limited judgments. And, yes, even their potential usefulness to us on that strange terrain, the future.

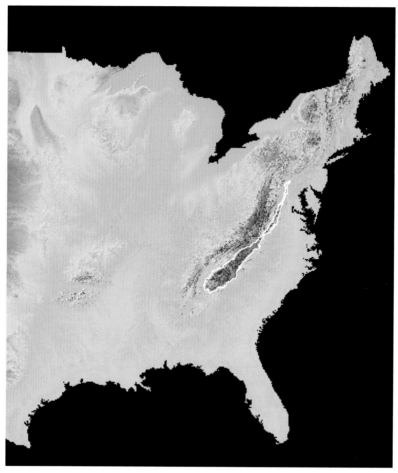

Plate 1. *The eastern United States, showing the outline of the Blue Ridge ecosystem. (Sources:* Keys et al., *Ecological Units of the Eastern United States*, CD-ROM; Hermann, *Southern Appalachian Assessment GIS Data Base*, CD-ROM)

Plate 2. *The Blue Ridge ecosystem and surrounding region.* (*Sources:* See plate 1)

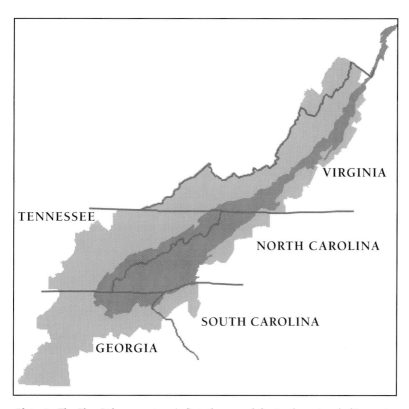

Plate 3. *The Blue Ridge ecosystem (red) is the core of the Southern Appalachian region* (outline of Southern Appalachian counties shown in green, state lines in gray). (*Sources:* See plate 1)

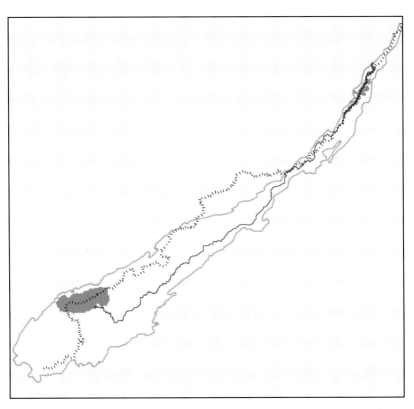

Plate 4. *Locations of the Blue Ridge Parkway and Skyline Drive (solid line), the Appalachian Trail (broken line), Shenandoah National Park (at the northern end), and Great Smoky Mountains National Park (in the southern section) in the Blue Ridge ecosystem. (Sources: See plate 1)*

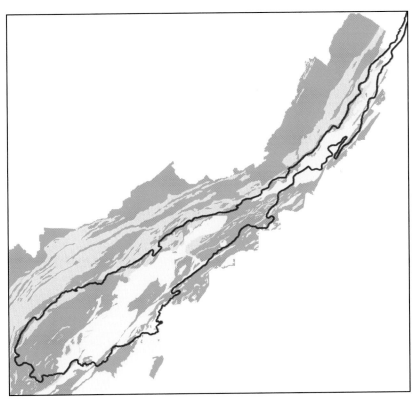

Plate 5. *Acid rain and Blue Ridge rocks.* Most of the geology of the Blue Ridge is sensitive to acidification by acid rain. The orange shade on the map indicates highly sensitive areas; yellow, somewhat sensitive areas; the lightest areas have low sensitivity. (*Sources:* See plate 1)

Plate 6. *Three views of a national park.* This is how our view of the landscape at Great Smoky Mountains National Park, and throughout the Blue Ridge, is affected by air pollution. These are computer simulations, based on 1988–95 observed data. The top view reproduces a clear summer day, but it includes natural or background haze; the middle view is an average summer day; the bottom view is a "worst day" pollution scene. Heavy pollution is common in summer. (*Source:* Air Resource Specialists, Inc., Ft. Collins, Colorado, using WinHaze 2.6.5 simulation program)

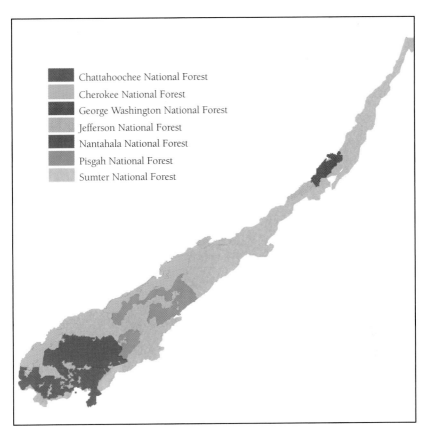

Legend:
- Chattahoochee National Forest
- Cherokee National Forest
- George Washington National Forest
- Jefferson National Forest
- Nantahala National Forest
- Pisgah National Forest
- Sumter National Forest

Plate 7. *National forestlands in the Blue Ridge. (Sources:* See plate 1)

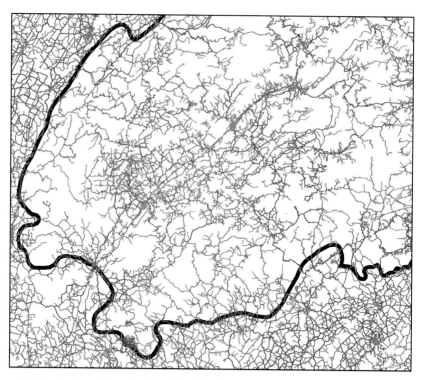

Plate 8. *Roads on the southern Blue Ridge, from highest to lowest traffic volume* (green, blue, red, brown). Some logging roads and other unimproved roads are not shown. A more detailed explanation appears with figures 13–16. (*Sources:* See plate 1)

Mountain lions, also called cougars or panthers, were last seen for certain in the Blue Ridge more than a hundred years ago. But they still reside, barely, in the realm of possibility. The nearest known population is in Texas, but seventeen cougar sightings were reported in Shenandoah National Park alone during one recent year.

A hiker from Alexandria may have seen one at dusk near White Oak Canyon:

"As the animal moved up the forested, rocky bank to my left," his letter to park officials reads, "it stopped and turned to look at me." Anticipating skepticism, he offered credentials: "I am an intelligence officer, an attorney, a Southern Baptist, and I wash behind my ears once a week."

Cougar experts say if one roams even a large area of the Blue Ridge, evidence of its presence should not be hard to find. But despite intensive searches and many reported sightings in the mountains over the years, no track, scat, hair, carcass, or photograph of a native cougar has so far been found.

Some of the sightings turn out to be bobcats, dogs, or captive cougars abandoned in the forest by weary and irresponsible owners. Several such "pets" have been confiscated by game wardens. The skull of a young African lion was found in the North Carolina Blue Ridge, and the bones of a cougar recovered in Pennsylvania were those of a Central American cat that had been raised in captivity.

A few of the phantom cats have been seen over the years by knowledgeable, experienced observers, however, including park rangers. "My gut feeling is that we have a breeding population in or around the park," says Shenandoah National Park ecologist Chip Harvey, though they may be released captives rather than wild in-migrants.

The disappearance of a predator like the mountain lion is an example of a human-caused ecosystem "disturbance," one of many created by our presence over the years. The flickering hope that a vanished species might still be afoot or abloom in some lost corner of the wilderness rouses us for many reasons, but for scientists it can be a sign that part of a damaged ecosystem may slowly be recovering.

Perhaps cougars, once abundant in the Blue Ridge, were a species unto themselves, or they may have been a variant of the western cougar. If they do not survive, some have suggested, why not reintroduce them, or a close relative from the western states?

When fish, bears, beavers, or eagles are returned to a former habitat, our own role in the natural landscape seems redeemed, our impacts reversible. But ecosystems aren't easily reassembled, and there may not be enough wild habitat left in the Blue Ridge to protect a population of cougars. There are plenty of deer to eat, wildlife ecologist Rainer Brock says. The problem is that the cougar habitat must be isolated as much as possible from the human species. Otherwise, the animals run afoul of cars and guns too often to survive. Cougars have also been known to attack and kill people. Though rare, it's a possibility that does not augur well for a long stay near us.

Peregrine falcons provide another example of the difficulty of reintroductions. They disappeared from the Blue Ridge and the rest of the East in the 1950s and 1960s, victims of the pesticide DDT. Reintroduction efforts in the Southeast were rewarded when the first mated pair of falcons returned to North Carolina's Linville Gorge in 1986.

But the odds are against tiny populations of birds, or any other organism, surviving natural pressures such as prolonged drought or disease. And falcons now face additional human-caused disturbance as well. In Virginia, populations of their natural enemies, the great horned owls, have increased with more intensive farming and urbanization in the region, which partly explains the very low falcon survival rates. By 1997, 323 peregrine falcons had been reintroduced in the Blue Ridge states, but only 11 nesting pairs had survived. In Virginia, only 1 nesting pair could be found, despite 121 attempted reintroductions over the years.

Elk, too, were hunted out of the Blue Ridge by the mid-1800s, but a small herd was transported from Yellowstone to the Peaks of Otter in Virginia around 1917. By 1935, eighty-five elk roamed there. Unfortunately, the experiment had ended by the early 1970s, when the last of them disappeared. Many died of "lead poisoning"—they were shot by farmers whose orchards supplied favored elk food. Others, unadapted to the diseases of a new ecosystem, succumbed to a brain parasite carried by local deer. Perhaps inbreeding also played a role. In any case, the elk population was not sustainable.

"Sustainability" is a leaden term, often complained of but not, so far, improved on. It poses the deceptively simple question, "How long can this go on?" In the Blue Ridge, sustainability means, among other things, the chances for long-term survival of a population of plants or animals: will they be able to withstand the challenges—natural or human-caused—that are certain to occur over time? Falcon, elk, and mountain lion populations without enough wild habitat, or some other guarantee of protection from human influences, are not sustainable.

As for red wolves in the Blue Ridge, the jury is still out and will remain so, indefinitely. The red wolf's range originally extended from the Atlantic to the Gulf coasts, as far north as Ohio and Pennsylvania, and west into Missouri. One source says that the last wolf in the Smokies was killed in 1887, though a map published in a federal report twenty years later still showed "big wolves" on the Tennessee–North Carolina border.

Fewer than a hundred red wolves survived by 1970, stranded in the coastal thickets of Louisiana and Texas, and they had begun to interbreed with coyotes. Then, at half past the eleventh hour, they were declared "endangered." The remaining few were captured by the U.S. Fish and Wildlife Service, bred with other captives, and conditioned for reintroduction.

The agency released a mated pair of red wolves in Great Smoky Mountains National Park as an experiment in 1991, and thirty-five more followed during the next five years. But at the end of that time only thirteen adults and an equal number of pups roamed the park. Many were recaptured, for a variety of reasons. Some died: two of antifreeze poisoning, one of a poacher's bullet, and the rest through natural or unknown causes.

The losses are "something we have come to expect," wildlife biologist Christopher Lucash, who supervises the Smokies project, says. "It's not that alarming to us to lose a high percentage of captive-raised wolves. Once you get a wild-born population going out there, they will do much better."

The wildlands of the Southern Appalachian region might well support a couple of hundred red wolves. By the middle of the next century, if all goes fairly well, they will no longer be an endangered species. There is little doubt that they can survive, given time and space. Their future has been shadowed, instead, by the uncertain commitment of the human population on which they ultimately rely. Much of the project's time and effort has been expended not in aiding the wolves but in fending off trouble between wolves and humans.

"These wolves are predatory animals and they have become an exceedingly dangerous presence in eastern North Carolina," North Carolina's Senator Jesse Helms once told fellow lawmakers. "They slink onto private property, they attack and feed upon farm animals and livestock, and we have reports that at least one child has been bitten by a red wolf. . . . It has become exceedingly dangerous to the people." Helms's recent campaign in Congress to kill funding for the red wolf program failed by only two votes.

Lucash and Gary Henry, the wolf program's administrator, say that, oddly, no one reported a "wolf attack" to local officials in North Carolina or to the Fish and Wildlife Service. One child in the mountain counties had reportedly been bitten by a big red doglike animal with a metal ear tag, in an area where stray dogs are common. Red wolves are not ear-tagged and have only hints of reddish hair on their pelts. There is no record of a healthy red wolf ever attacking a human anywhere.

"Just the word 'wolf' will cause some people to panic," Lucash says. "They worried that we were going to bring in a couple of semi trucks full of bloodthirsty killers and turn them loose in the park—a wedge of death, coming down out of the mountains. After a couple of years, when nothing happens—there are still plenty of deer, raccoons, and rabbits, and of course no attacks on humans—it's no big deal."

The wolf reintroduction program has elaborately accommodated the concerns of livestock producers. Healthy cattle can easily run off red wolves, but newborns and sick animals can't, so producers were promised that they would be fully compensated for any confirmed killings. Any wolves leaving the confines of the national park would be retrieved. Rules against shooting wolves if they were caught in the act of killing livestock were relaxed, despite the wolves' extreme rarity and the twenty years of work and expense invested in their survival. (There are about 300 red wolves, and 104 million cattle, in the United States.)

The sponsoring agencies were happy for the opportunity to placate, however. All reintroductions, especially of predators, are test cases. They will accumulate either ill will or good among humans—now the key element of sustainability. In the future, Lucash hopes, the humans will become acculturated. Then the wolves will "belong" in the landscape again, like bobcats, black bears, or beavers, and they will be allowed to disperse. The challenge for the wolves' patrons is to work with landowners to try to get them to realize the long-term benefits of reintroduction.

Discreet inquiries have been made among federal and state officials about chances for red wolf reintroduction farther north in the Blue Ridge, but the map shows few opportunities. There is suitable habitat but, for now, insufficient enthusiasm. National forest administrators, beleaguered by conflicts over logging, may be sympathetic to the wolf program, but they are unable to respond to local objections. "They don't need more controversy," Lucash says. "They're busy. They've got other headaches."

"We may be looking for a place instead of the Southern Appalachians," he adds. "This may not work. We admit that we could just up and find we are not going to get a population established here, for whatever reason. It probably is going to be politics more than anything. Part of the problem politically here is that we're going to need more than the park. We're interested in the Cherokee, Nantahala and Pisgah national forests, and even the Chattahoochee in Georgia."

Support for the wolves would have to be plainly evident among state officials for such a project to be feasible, even to propose. "It would have to be an overall adoption by the state, not just the federal government, to allow wolves into those areas," Lucash says. "State officials respond to local constituents. So it's a long process of education, and flirtation, and proposing."

The National Research Council, asked by Congress to study endangered species programs, concluded that preservation of habitat is the major factor in success or failure. The council recommended that:

- Emergency "survival habitat" should be provided for some endangered species immediately, without regard to economic impact.
- Federal agencies (and the rest of us) should think in terms of saving and restoring whole ecosystems rather than one species at a time.
- Current laws cannot by themselves prevent more extinctions.

In the Blue Ridge, the lesson of the falcons, wolves, mountain lions, and elk is the same. Threatened species can't be saved, or reintroduced, without saving enough wild, protected living space.

And the missing predators have essential work to do in making an ecosystem sustainable. They eat meat. In their absence, populations of their prey can explode, and the ecosystem lurches out of balance.

Deer, for instance, were almost hunted out of the Shenandoah area by the time the national park was established in the 1930s. But the absence of both animal and human predators, and the availability of food from nearby agricultural areas during lean winters, allowed deer populations to increase. Density in some northern sections of the park is high now, averaging 78 to 104 deer per square mile.

The deer eat nearly everything, including the seedlings of both hardwoods and conifers, grass, shrubs, and acorns. Where there is high deer density, some places are stripped bare. "There is nothing in the understory. It's like a moonscape," Smithsonian research scientist Bill McShea says.

As the diversity of plant life plummets under the pressure of a deer herd freed from predators, the variety of animal species drops off, too. Vegetation dictates which animals can survive on the land, and the deer now dictate the vegetation.

Kentucky warblers that used to be common in McShea's research area are nearly gone. "Over the last 15 years the deer have pretty much eaten the place out and the warblers can only nest in the fenced 'exclosures' we have set up, or in bottomland hardwoods, where things grow faster than the deer can browse," McShea says.

By contrast, Kentucky warblers, American redstarts, and hooded warblers have multiplied within the fenced-off "exclosures." So have small mammals such as white-footed mice, gray squirrels, wood rats, and chipmunks. They compete with deer—as do grouse, turkey, bear, and other animals—for the protein-rich annual acorn crop.

The researchers predict that in fifty years, if the deer population remains dense, the forest floor will be open and parklike with an aging group of taller trees and a near-total absence of new, young trees and shrubs. The birds and insects that depend on the understory will also be missing.

In places such as the wetlands in the Big Meadows area of the park,

SOLUTIONS 1

National Parks Face the Next Century

Two of the three National Park Service units in the Blue Ridge—Great Smoky Mountains National Park and the Blue Ridge Parkway—are the most frequently visited in the United States, and Shenandoah National Park has an additional 1.7 million visitors each year.

Parks personnel are government employees, so they don't talk much with the visiting public about funding cuts, badly eroded trails, pollution, or political pressure from local communities. But a succession of reports have called attention to these problems, as well as declining ecosystems, deteriorating facilities, and an impoverished research effort in the parks. Many employees report feeling a lack of public and political support for their work. A recent announcement for a professional meeting of park scientists and administrators gives a sense of their outlook on the future:

"While the bedrock assumption underlying the creation of parks and reserves is that they will be protected in perpetuity, today's world is characterized by the dizzying pace of technological change, rapid human population growth, large-scale alteration of ecosystems, the disintegration of shared cultural views of history, declining government budgets, and an increasingly fragmented and volatile political climate."

A recent Park Service publication on its own future noted that the agency "has a phenomenally dedicated work force, some of the nation's most treasured resources under its management, and widespread support from the American public. At the same time, however, it suffers from declining morale . . . serious fiscal constraints, and personnel and organization structures that often impede its performance."

A steering committee that drew on recommendations from seven hundred national parks personnel produced advice for the future. Here, edited and paraphrased, is some that applies to the Blue Ridge:

- Despite repeated calls for a strong research program over the last three decades, the Park Service's response has been sporadic and in-

hungry deer also threaten the survival of populations of rare and endangered plants, biologists say. Deer have browsed special experimental plantings of native trees, even when the trees were sprayed with bitter-tasting deer repellents.

Deer dominate plant life in sections of Great Smoky Mountains National Park such as Cade's Cove, too. Rare plant specialist Janet Rock says they have injured small populations of some of the rarer plants. The single specimen of the purple fringeless orchid in the park grows in a place where deer bed down for the night. "Deer think of orchids as ice cream," Rock consistent. In a world where ecological management has become a primary concern, this is a serious deficiency.

- In natural parks (as opposed to historical ones), the central task of the Park Service is to maintain, within a world mostly governed by human activity, a substantial sample of functioning natural systems not dominated by humans. Custodial management and the promotion of tourism were once sufficient, but now scientific expertise and first-class research capability have become crucial.
- Parks were once isolated from most human activity, but they are now inundated by external threats and need to be managed in partnership with those who share park resources and boundaries.
- The Park Service must continue to strengthen access to the parks and efforts to educate the public about their value.
- In those instances in which the ecological balance is under threat or is uniquely fragile, people should be kept away.
- The service must have the capacity to respond to threats, whether they come from a dam at the park boundary, air pollution from a facility 100 miles away, or global warming.
- The boundaries of many formerly remote natural area parks are increasingly suburbanized. This is often spurred by state and local governments anxious to capitalize on tourism-led regional growth. This rapid change in the areas around and near many parks raises concerns about externally generated degradation in the parks themselves.
- The Park Service should plan for innovative transportation systems in the parks that deemphasize the use of automobiles.

Sources: National Park Service, *Park Science* 16, no. 4 (Fall 1996): 31; National Park Service, *National Parks for the Twenty-first Century*, 2, 17–18, 105–6, 86.

says. "They always seem to like the rarest of plants." She calls them locusts with hooves.

But the National Park Service has learned that programs to reduce animal populations can provoke fierce public opposition. The agency's mission to preserve its holdings "unimpaired for future generations" often collides with other park values, such as having plenty of deer for visitors to see.

John Rappole also works on the Smithsonian's research on ecosystem disturbances caused by deer. "It really comes down to what resources the national park wants to protect," he says. "Do they want the high deer populations or do they want a relatively natural ecosystem?"

Devastation can come to the mountains in wholly natural disturbances, of course. One of those arrived on June 27 and 28, 1995, after five straight days of rain. During those two additional days, a slow-moving storm stalled against the eastern front of the Blue Ridge in Virginia, held in place by light winds.

In the mountains, prolonged, intense rain engorged streams and loosed violent mud flows, debris flows, and landslides as soils liquefied. During a two-hour period around noon on the 27th, 12 inches of rain fell at some locations. At Graves Mill, just east of Shenandoah National Park, 24 inches of rain reportedly fell in a day.

Mud, boulders, and trees were hurled down the Staunton, the Rapidan, and the North Fork of Moormans River within the park, scouring out streambeds and the vegetation on what had been the banks. West of the hamlet of Wolftown today it looks as if a giant has raked its fingernails straight down from the tops of the ridges, leaving parallel scars all the way to the bottom. The quiet, narrow, tree-shaded stream that once was the Staunton River is now a barren, hugely bouldered, 200-foot-wide "blown out" channel, punctuated by pileups of hundreds of logs blasted clean of bark.

Insect, salamander, and fish populations were erased from some streams. Three weeks after the flood, no fish of any kind were detected in a section of the Staunton that had harbored eels, dace, suckers, and as many as 83 brook trout in prior years. In the North Fork of Moormans River, where 824 fish of thirteen species had been tallied just before the flood in mid-June, a total of 6 fish were found on July 10.

But a year later brook trout had returned to the upper and lower margins of the most severely affected section of the Staunton. Fishery research biologist C. Andrew Dolloff of Virginia Tech found trout "almost every place we looked—they're back in there already." Eels, dace, suckers, and salamanders had also returned. "It was an eye-opener for me in terms of the resiliency of fish populations," Dolloff says.

Such storms batter one locale or another in the region every ten or twenty years, on average. In fact, mountain ecosystems can't be healthy without periodic death and destruction. Wildfire, ice storms, raging rivers, waves of hungry insects, and howling, uprooting winds are all part of the natural scene.

So what basis do we have for accepting the natural disturbances and worrying about the human-made ones? It's a question, scientists say, of their intensity, scale, and frequency. There's a difference between a haircut and a decapitation; between falling down stairs once and doing it on a weekly basis; between catching a cold and catching tuberculosis.

In our era, the human disturbances are paramount. The extermination of predators, the construction of dams, alien plant and insect invasions, thousands of tracts of land cleared each year for houses, stores, and roads— these are often more widespread, as well as more frequent, more sudden, and more permanent. They compound the effects of both natural disturbances and other, human ones.

Other than nuclear war, global warming caused by the "greenhouse effect" is the broadest-scale human disturbance we now contemplate. It is the result of several kinds of atmospheric gases that trap solar heat within the earth's atmosphere, just as the glass in your car admits the sun's rays but prevents heat from escaping. Human activities are causing a build-up of those gases in the atmosphere.

A few climate scientists may remain unconvinced that global warming is in the offing, and of its destructive potential. But a United Nations panel and an international group of scientists have reached a different conclusion. They advise us to turn off, as much as possible, the industrial spigot through which we pour "greenhouse gases" into the atmosphere, because the prospect of warming is real.

These scientists can't even picture the future climate of the whole Northern Hemisphere with much certainty, let alone the Blue Ridge. They will, though, hazard some general statements: on average worldwide, warming of from 1.8 to 6.3 degrees Fahrenheit may occur over the next one hundred years.

The earth's climate has shifted countless times in the past, and we still have evidence of some of the changes that took place during the more recent episodes. Fossilized pollen tells us that when the climate gradually warmed and the glaciers began to recede about sixteen thousand years ago, the more cold-adapted plant and animal species had to "migrate" northward if they were to survive. Jack pines, common in the Blue Ridge then, are now common in Ontario, Canada, instead.

Tree species migrate as fast as their methods of seed dispersal, soils, moisture, and other factors will allow. Eastern spruce populations followed the retreating glaciers north at a pace of about 750 feet a year; pines averaged a fraction of that. But the predicted "greenhouse" climate change will be distinctly different from the changes that occurred then.

The same amount of warming that took thousands of years as the last ice age ended may occur in a single century of our newly improvised climate—

SOLUTIONS 2

Climate Change on the Clock

How long will it be before the industrial facilities that emit the greenhouse gases that contribute to global warming are replaced under typical economic conditions?
Years to decades, unless they are required to close earlier.

If emissions of greenhouse gases stop increasing, how long will it be until concentrations of those gases are stabilized in the atmosphere?
Decades to millennia, for the longest-lived gases.

If concentrations of greenhouse gases in the atmosphere are stabilized, how long will it take for the climate to reach equilibrium?
Decades to centuries.

If the climate is stabilized, how much time will pass before sea levels reach equilibrium?
Centuries.

How much time will pass before damaged or disturbed ecosystems can be restored or rehabilitated?
Decades to centuries, if it can be done at all: some changes such as extinctions are irreversible, and it may be impossible to reconstruct or reestablish some ecosystems.

Source: Watson et al., Climate Change 1995, 4.

the coming one. It is unlikely that all of the species that are part of Blue Ridge ecosystems could move and adapt so quickly. "Entire forest types may disappear, and new ecosystems may take their places," the UN report states.

If they do survive the pace of this forced march to the north, Blue Ridge species must also find ways to leapfrog vast biological roadblocks on the way. Fragmentation of forests for logging, agriculture, and urbanization "is rapidly removing the migration corridors that existed for forest species during earlier periods of climatic change," one set of authors warns. Without a network of nature preserves to allow migration, global warming may drive some of these trapped species into extinction.

In the flatlands of the southeastern United States, the predicted greenhouse warming could shift the temperature lines on the weather map to the north, 90 to 340 miles. In the year 2100, then, average temperatures in Charlottesville, Virginia, could be comparable to those in places like Danville, Virginia, or Columbia, South Carolina, today.

But the effects would be much more intense in the mountains. As you climb, average temperatures cool by 2 or 3 degrees Fahrenheit for every 1,000 feet of elevation. Global warming in just the coming century could, in effect, lop 500 to 1,800 feet off the mountains in terms of average temperatures. Higher or more frequent "spikes" in summer temperatures would bring even sharper stresses.

Such change could destroy what is left of already rare high-elevation habitat in the Blue Ridge. "Some species with climatic ranges limited to mountain tops could become extinct because of disappearance of habitat or reduced migration potential," the UN report says. "Recreational industries, of increasing economic importance to many regions, also are likely to be disrupted."

Climatologists underline the uncertainties in trying to predict warming effects. They also stress the necessity of attempting it, since decisions must be made now to affect climate trends fifty years or more in the future. Climate shifts will be incremental rather than explosive. Unless we bring ourselves to focus on such changes, they are not always visible within the time scales that normally occupy our attention.

In his book *The Next One Hundred Years*, author Jonathan Weiner evokes the problem. "We don't respond to processes," he writes. "We respond to events." And he tells this story: a biology teacher dropped a frog into boiling water. The frog jumped out, fast. Then the teacher put the frog in a beaker of cool water and turned on a burner underneath, warming the water gradually. The frog kept swimming around until it boiled to death.

The highest places in the Blue Ridge are a gray-green archipelago—islands of spruce-fir forest, isolated by the passage of millennia. These were once a broad dominion covering, during a colder time more than a hundred centuries ago, 700,000 square miles from Missouri to the Carolina Piedmont. As the climate warmed, the cold-adapted spruce-fir ecosystem retreated to a scatter of high peaks and ridges from Mount Rogers south to Great Smoky Mountains National Park (fig. 3).

Today, only 103 square miles of this type of forest persists in the Southeast, more than three-fourths of it within the park. The spruce-fir climate zone, similar to parts of Canada, has generated a unique habitat that is rich in rare plants and animals. During the lonelier seasons at the highest elevations, temperatures drop as low as −34 degrees Fahrenheit, which the wind-chill factor pulls down toward 100 below. The winds have been measured at a brisk 175 miles an hour, and 103 inches of snow falls, on average, each year. Eight plant species, including the Fraser fir, are found only in this ecosystem.

Among these patches of boreal forest, Mount Mitchell has always been an icon, because it is the highest peak east of the Rockies. But now, there and throughout the Blue Ridge, all that remains of tens of thousands of mature Fraser firs is a ragged palisade of bleached, skeletal trunks. As they perish, the ecosystem they support also declines.

Mount Mitchell's recent history, tightly interwoven with our own, explains what has befallen the high forests. By 1915 a logging railroad reached almost to its 6,685-foot summit. While the logging was under way, North Carolina governor Locke Craig, appalled by the devastation, led a public movement to create Mount Mitchell State Park. "It was on a place like this that Moses communed with God, who revealed himself to man," Craig declared. "He has given this sacred place to us, and we should do our best to preserve it . . . forever."

A small part what is now Mount Mitchell State Park escaped logging, but square miles of surrounding slopes were denuded. Light, strong spruce wood was especially suited for the airframes of World War I biplanes. It was so much in demand that the U.S. Army sent two hundred soldiers to the area to help the Suncrest Lumber Company with its cut.

Intense fires, often started by sparks from logging locomotives, were common in the dried-out slash debris that remained after the cut. The organic matter in the soil was incinerated, and some fires jumped into standing timber nearby. When the era ended, this already rare kind of forest had been reduced by half. Much of the land—cut over, burned over, and eroded—was unable to regenerate spruce and fir. On some parts of Mount Mitchell no trees of any kind have returned after the passage of eighty years.

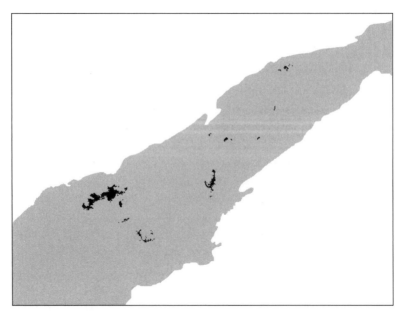

Figure 3. *Spruce-fir ecosystem "islands" in the Blue Ridge.* Mount Rogers is the northernmost patch, and the largest is within Great Smoky Mountains National Park, farther south. (*Sources*: Keys et al., *Ecological Units of the Eastern United States*, CD-ROM; Hermann, *Southern Appalachian Assessment GIS Data Base*, CD-ROM)

In the aftermath of the logging, erosion choked the mountain streams with silt, threatening water supplies for cities such as Asheville. The U.S. Forest Service was under intense public pressure to try to reestablish trees on the barren mountain slopes. As part of an experimental reforestation project, the agency transplanted non-native species, including European firs, to Mount Mitchell in the early 1930s. They carried with them a tiny insect native to Europe, the balsam woolly adelgid, which was discovered on Mount Mitchell in 1957. By then, the whole mountain was infested, and the insect was riding air currents to the rest of the Blue Ridge.

In one of several life stages the adelgid, about the size of a period on this page, sprouts curling, white, woolly-looking waxy threads. Inserting a tube into the bark, it feeds on sap and secretes saliva, whose chemistry disrupts and hardens outer wood cells. Death comes to the fir, by a kind of strangulation, in two to seven years. An estimated 90 percent of Fraser firs in Great Smoky Mountains National Park had been wiped out by the late 1980s, and it was suggested as a candidate for the federal endangered species list.

Handsome, fragrant young firs arise from these graveyards, and they may in turn produce new seed, but future generations will succumb to the adelgid after twenty years or less. The survival of even these forever-young stands is in doubt, because seed production in fir trees is very low.

Whether or not individual trees persist in some form, the ecosystem of which they were a key component is already severely damaged. Some of the changes are evident along the trail over Mount Collins, in Great Smoky Mountains National Park. The dense overhead canopy that used to provide continuous shade and wind protection is now just fragments against open sky. The few remaining red spruce trees are often felled by ice and wind, breakage and blowdown, without the protection of the fir. The soils, now exposed to the sun and wind, have become hotter and drier, so shade-loving plants die off and thick, tangled undergrowth, especially blackberry, springs up.

Biologist Kerry Rabenold monitored the changes on Mount Collins over two decades spanning the adelgid's arrival. During that period, 95 percent of the mature firs died. No longer protected by neighboring firs, two-thirds of the spruce trees succumbed to windthrow, along with half of the less common yellow birch. In all, fewer than one-sixth of the trees of any kind remained.

Populations of the eleven different species of breeding birds that frequent this kind of forest, such as solitary vireos, red-breasted nuthatches, black-throated green warblers, and golden-crowned kinglets, became increasingly scarce. The sum of birds of all species dropped by half, and the trend pointed to further declines.

Predictably, other species have been affected, too. The loss of fungi, lichens, and seeds in what remains of the spruce-fir habitat may have displaced populations of endangered northern flying squirrels, with still-to-be-reckoned consequences for their numbers.

The reckoning is almost complete, however, for the spruce-fir moss spider, an animal about as large as the head of a pin whose lifeways puzzle the few scientists that have observed it closely. One curiosity is that its only close relative lives a continent away, in the Cascade Mountains of the Pacific Northwest. These two species are the northernmost examples of the primitive tarantula family, which is found almost exclusively in the tropics.

Most spiders live only a single season, but it is possible that this type of tarantula can survive for as many as ten to forty years, despite the forbidding extremes of the high-elevation climate. Also a mystery is its habit of living in groups that include all age classes; it seems unable to survive as an individual.

"They are unusual, intriguing. Only a handful of people have ever even seen one," arachnologist Joel Harp says. "There's just so little known about this species." Hundreds were seen on Clingman's Dome in Great Smoky Mountains National Park as recently as 1980, but a resurvey a few years later found that all had vanished. They had inhabited damp mats of moss that grew on the rocks beneath the Fraser firs. When the firs died, sun and wind parched and killed the mosses. An urgent search along the rest of the Blue Ridge has so far located only three places where the spiders remain. Attempts to breed them in captivity failed and were discontinued. The moss spider joined the endangered species list in 1995.

In Europe, where the balsam woolly adelgid is a native rather than an exotic, trees have natural defenses and are only mildly affected. Firs there have co-evolved with the insect over countless thousands of years. In the Blue Ridge, the Fraser fir has no time to adapt.

And in the adelgid's original range other insects, fungi, bacteria, and viruses have also evolved to prey on or compete with it, keeping its numbers in check and further limiting its role. That kind of balance in populations is a characteristic of healthy ecosystems, but the adelgid has no effective predators here.

Most species of all kinds are relatively stable in another way, too. Though their ranges naturally grow and contract a bit over decades or centuries, they rarely move thousands of miles or cross barriers such as mountains or oceans on their own. But today, the movement of species into new areas is common—so common that it "dwarfs" natural rates of change, researchers say.

This phenomenon, sometimes called biological pollution or biopollution, is the result of our own movements and of our prolonged inattention. A few purposely introduced species, such as food crops, have been hugely beneficial, but nearly all introductions these days are inadvertent. By the mid-1950s, ecologist Charles Elton had already carefully redefined the word "explosion" as a technical term for "terrific dislocations in nature. We are seeing huge changes in the natural population balance of the world. . . . We must make no mistake. We are seeing one of the great historical convulsions in the world's flora and fauna."

The sudden fast-forward in biopollution is also boosted by the fact that, in general, non-native species succeed best in ecosystems that are already disturbed by human activity.

Many of these "exotics"—meaning species that are out of place, alien, beyond their natural range—die off or pose no threat to native biodiversity when they arrive in North America. They hit a brick wall here: the climate, the vegetation, competing organisms, or some other factor. But other organisms, such as the adelgid, find paradise: an ecosystem with plenty of hosts and few or no genetic defenses, competitors, diseases, or natural predators. Then their populations leap.

It's hard to know in advance how an exotic will behave when we bring it to North America. As a study by the federal Office of Technology Assessment* puts it, "The movement of plants, animals, and microbes beyond their natural range is much like a game of biological roulette."

More like Russian roulette, though. Introductions of exotic species are almost entirely in human hands. The self-inflicted damage they incur in the United States is staggering, whether we measure in terms of biodiversity or in dollars: hundreds of millions to several billions of dollars in losses, annually, to forests, rangelands, agriculture, fisheries, and industry. We have pulled the trigger more frequently during the past century, as the pace of international travel and trade quickens.

Plants, animals, insects, and diseases moved with comparative ease throughout the Northern Hemisphere during an earlier time, too, 1.1 billion to 250 million years ago when the Blue Ridge was created. Then, the continent-sized moving plates that later became northern Europe and Africa collided three times into what is now the coast of eastern North America. These events created Himalayan-scale, or perhaps Rocky Mountain–scale, wrinkles in the earth's crust. Worn down as time passed, they are now the Blue Ridge and the rest of the Appalachians.

Then the Atlantic gradually opened, and the evolution of plants and animals in North America proceeded in increasing isolation. The last eastward land bridges linking them to organisms on land masses to the east probably disappeared during the early Tertiary, sixty-five million to forty-five million years ago. By now, species here are significantly different from their relatives on the other two continents.

Humans arrived in the Southeast at least as early as twelve thousand years ago, doubtless bringing along a few exotics such as favored food or medicinal plants. Among this suite of arrivistes, then, the humans had served as the carrier, or vector, for the introduction of the other species. When Europeans came, the pace quickened. By 1800, perhaps twenty-eight kinds of insects had been introduced into the United States. By 1980, there were more than fifteen hundred.

The most recent count of free-living exotic plants, land vertebrates, insects, spiders, fish, mollusks, and plant diseases in the United States found 4,542 of them. More than 200 of these exotics were introduced or first detected in the United States just between 1980 and 1993. Fifty-nine of those are known to be harmful. Humans now easily qualify as the chief vector for new biopollution—exotic diseases, plants, insects, and animals—that is introduced to our native ecosystems.

One result is that those islands on the map—the spruce-fir enclaves, the

*The Office of Technology Assessment was abolished by Congress in 1996.

national parks, and the entire Blue Ridge itself—only resemble islands. In terms of our defenses against the rising tide of exotics, the real borders of the Blue Ridge ecosystem extend to Wilmington, Richmond, the Great Lakes, Brooklyn, and beyond.

Richmond has a fine municipal park, Maymont, which at the turn of the century was the estate of Major James Dooley. An avid collector of plants from around the world, he may have acquired a Japanese hemlock to ornament his ample grounds. Two species of hemlock are native to Japan, and Dooley's specimen could have arrived with an infestation of hemlock woolly adelgid, an Asian relative of the balsam woolly adelgid. We may never know precisely how it made its way here.

But around 1953, the hemlock woolly adelgid was discovered on Eastern hemlocks at Maymont Park. Eastern and Carolina hemlocks are the species native to, and common in, the Blue Ridge. The adelgid was regarded as a minor pest, but within a few years, infestations were discovered to the north and west. Then the adelgid reached natural stands of hemlock in the Blue Ridge, and its behavior changed sharply. Severe damage, high mortality and a faster rate of spread were reported.

At its current rate—10 to 15 miles each year—the insect will have infested all of the Blue Ridge by around the year 2015. In its path are some of the largest old-growth Eastern hemlock stands left in the world, in Great Smoky Mountains National Park. The adelgid was found as far south as North Carolina by the late 1990s.

By then, an estimated 20 percent of the hemlocks in Shenandoah National Park were dead, and fewer than a third were still in fair health. "The results and future are grave," park exotics specialist James Åkerson says. The cottony white stuff on needles that indicates the adelgid's presence could be seen all over the northern Blue Ridge, from the park's famed Limberlost grove of old-growth hemlocks to a remote ravine along the Piney River where one dead hemlock, bared of its bark by woodpeckers, looked flayed—a huge column of vivid coral.

"There are trees that seem to have survived infestation," says Virginia state forest health specialist Tim Tigner. "I have even seen a very few trees that have recovered from a very sad state to one of apparent good health. But we can't really say what's going to happen down the road. Most trees continue to die even when the adelgids disappear."

Research entomologist Mark McClure says that Eastern hemlocks probably will not become extinct. The Carolina hemlock may, however. It is found only in the Blue Ridge, and its populations and distribution are much smaller.

Despite the speed of its current exit from the ecosystem, the fossil pollen record suggests that Eastern hemlock first moved into the Southern Appalachians some twenty million years ago. The species has some extraordinary properties. It grows slowly and can remain nearly dormant over many decades as a smaller understory tree, awaiting a chance disturbance—a storm, the death of a neighboring tree—that will bring sunlight and competitive advantage. When Europeans arrived here, they found forests of big hemlocks, sometimes 100 to 160 feet tall and more than five hundred years old, growing from Nova Scotia to Alabama.

"A hemlock stand tends to create an environment that is suitable for perpetuating itself," forest entomologist John Quimby writes. "There are no darker, cooler places in the forest than under a hemlock canopy." Soils tend not to dry out in such conditions, and a wide variety of herbs and shrubs thrive there as ground cover.

Brook trout, too, are more common in streams cooled under this deep shade. Hemlock is long lived and extremely tolerant of shade, so it can outlast other species until it begins to dominate a stand of trees. In the Northeast, hemlock has been found to provide food, nesting or roosting sites, or winter cover for ninety species of birds. Three probably depend on it exclusively and a dozen others "primarily" or "substantially." About 14 percent of mature Blue Ridge forests at lower elevations may be made up of hemlock.

Scientists expect that as hemlock fades from the forest, streams will warm, populations of brook trout will decline, and the exotic brown and rainbow trout may increase; diversity in some classes of plants will diminish as ferns and stream algae become more common; invasive exotic plants will increase; and trees such as yellow birch and red maple will replace the hemlocks.

"Aesthetically, the hemlock has no equal in the East," Quimby writes, though partisans of other trees—dogwood, for instance—might dispute the point.

Forest Service plant pathologist Robert Anderson was the first to discover anthracnose fungus on dogwoods in the Blue Ridge. He had been called in to investigate ailing trees deep in the wilds of northern Georgia in 1987. Worried about taking a sample back to his laboratory in Asheville, he sent it directly to a lab in Connecticut for analysis and identification. Anthracnose, probably introduced on exotic nursery stock, had already been laying waste to dogwood populations in the Northeast, but there was some hope that climate factors might keep it out of the South.

In the few years since, Anderson has witnessed the explosion of the fungus throughout the entire range of dogwood in the Southern Appalachians and helped track the result. Above 3,000 feet most dogwoods are dead, and the rest will be gone shortly after the turn of the century, he says. Between 2,000 and 3,000 feet all the trees in moist, cool, or shaded areas, or on north-facing slopes, will also die off, but sunlit areas will be okay. Perhaps 30 percent of the dogwoods will remain at those elevations. Below 2,000 feet, only the dogwoods in very wet, cool, shaded areas will die. When the fungus has reached its limit of infection shortly after the turn of the century, it may have claimed as many as half of all the dogwoods that grew on the Blue Ridge. "We've seen the loss of millions of them," Anderson says.

Dozens of bird species as well as bear, turkey, and deer make use of the tree. Dogwood also has the valuable property of removing calcium from the soil and concentrating it. As its leaves drop to the ground and decay, more calcium, vital for the health of most kinds of trees, is made available to the ecosystem.

Another, earlier victim of a wave of exotic fungal infection, the chestnut, is still common as a small, shrubby tree in the Blue Ridge. It resprouts from the roots of long-dead giants but seldom reaches more than a few feet of growth until it is knocked down again by chestnut blight. Dead dogwoods, however, do not resprout.

The chestnut blight fungus was first seen in Brooklyn—which, at the time, must have seemed a long way from the Blue Ridge—in 1904, probably introduced on nursery stock from Asia. Its bright orange spores leaped through the eastern forest like flaming cinders and destroyed as many as a billion chestnut trees over 224 million acres.

Chestnuts were so plentiful that they made up about a quarter of the tree cover, and pure stands were not uncommon in the Blue Ridge. The chestnut's rot- and insect-resistant wood supplied a long list of human uses, from fine furniture to mining timbers. It was the most economically important tree in the East, and it was able to grow in relatively poor soils. The blight had spread throughout the Blue Ridge by the mid-1920s, and the trees were nearly all dead two decades later. The sudden obliteration has been called one of the greatest changes in a forest ecosystem ever recorded.

Five insect species that we know of disappeared with the chestnut forest. Its reliably heavy annual production of protein-rich nuts had sustained many kinds of wildlife. As the trees sickened, dropped their leaves, died, rotted, and toppled, oaks took their place. Oak acorns are nutritious but far lighter fare for wildlife than chestnuts. And unlike chestnuts, the acorn crop fluctuates from year to year, sometimes sharply.

Vast numbers of new oaks were not well adapted to the former domain of the chestnut. They are now in ill health and not regenerating well over much of the Blue Ridge, a condition known as "oak decline." As oaks age in these less favorable sites, they are susceptible to root rot, drought, and, especially, the European gypsy moth, an introduced "defoliator" insect that prefers oak. Successive years of gypsy moth infestation kill a high percentage of the trees.

The European gypsy moth escaped from a failed attempt to produce hybrid silkworms in Massachusetts in 1869. Since then, it has become a champion among destructive pests and is now the dominant spring leaf-eating insect in eastern oak forests. The U.S. Department of Agriculture estimated losses of $764 million from the gypsy moth in one single year's outbreak.

The European gypsy moth had spread south as far as Shenandoah National Park by 1981. A couple of years after its first appearance there, you could hike long stretches in places like Nicholson Hollow, in the central section of the park, under an endless hail of tiny leaf fragments and caterpillar droppings and enveloped in a faint, eerie, rustle—the oceanic chorus of tens of thousands of chewing insects.

When the moth arrived, about 60 percent of the forest within the park boundaries was judged highly susceptible to defoliation. An unknown, but undoubtedly large, number of trees are now dead. "Ghost forests" of standing dead oaks can be seen throughout the park and massed along the higher ridges for dozens of miles to the south. It is not uncommon during hikes in the backcountry to hear the sudden pop and crash of one of these rotted trunks falling somewhere nearby.

The moth's "infestation front" has continued south, reaching as far as Montebello, Virginia, by the late 1990s, and spot outbreaks, quickly eradicated, have occurred in North Carolina and northern Georgia. The Forest Service has predicted that the moth will have infested all of the Blue Ridge by the year 2020.

That's not to say that oaks are finished, however. Gypsy moths have been in New England for 130 years, Cornell University insect pathologist Ann Hajek points out. "New England still has lots of oak and oak is the gypsy moth's favorite tree, so now oak is a smaller percentage of the forests than it used to be," she summarizes.

And even this formidable European import has some limitations. It munches on the leaves of oaks and a few hardwoods. That's a fairly restricted appetite compared with, say, the Asian gypsy moth, which is larger, has a much broader appetite, and grows more quickly. Fortunately, the Asian strain does not, so far, infest North America.

SOLUTIONS 3

A Fighting Chance against Exotics

The Office of Technology Assessment presented several options for change after an exhaustive study of the economic and ecological threats posed by exotics. Here are some:

- The most fundamental issue is whether the United States needs a more stringent and comprehensive national policy on harmful exotics. There is general agreement that the United States has no such policy now.
- The nursery, pet, fish farming, and agriculture industries can be expected to be "cautious about any congressional action that would make U.S. policy more stringent."
- Existing laws have significant gaps. One presidential Executive Order on controlling exotics issued in the 1970s has still not been fully implemented. The U.S. system for dealing with harmful exotics involves a "complex interplay" of agencies whose roles are ill defined.
- Federal and state agencies lack authority to deal with the invasion of exotics. Government and scientists don't have enough information on exotics to know how to respond to the threat.
- The National Environmental Policy Act could be applied to proposed releases of exotics, requiring a full environmental impact statement.
- New Zealand's approach includes detailed national standards regulating imports and strong authority to require bonds to cover potential costs of escape; a "user pays" approach to cover costs of inspection, surveillance, scientific analysis, and enforcement against violators; strict inspections of arriving passengers, baggage, and goods, with random checks to see if the system works; 100 percent treatment of arriving aircraft with insecticide; computerized tracking of imports, from arrival to unloading; detailed surveillance for forest pests; detailed planning for emergencies; extensive enlistment of public support for pest surveys and monitoring.
- The U.S. government currently places only a few piecemeal restraints on the importation of fish and wildlife and their diseases. Tens of thousands of different species (most of the world's fauna including insects) can be legally imported into the United States.
- States prohibit relatively few injurious species. Their standards for predicting harm are low, and enforcement is weak.
- Taken together, these gaps in federal and state powers and their lethargic responses are serious threats to the nation's ability to exclude, limit, and rapidly control exotics.

Congress could:

- Expand the list of banned exotics and speed the process for adding new ones. It took two years for the government to decide that zebra mussels were injurious. In that time, a precious opportunity was lost to control the spread of what is now a national aquatic menace.

- Tax the sale of exotic trees, shrubs, and flowers. Require licenses and license fees for plant imports.
- Require that the twenty federal agencies involved with exotic species regulations develop broad-based environmental education programs to increase public awareness of the threat, and require airlines, port authorities, and importers to intensify public education efforts about the threat.
- Ensure that all agencies have adequate authority and funds to handle emergencies.
- Pass tougher laws punishing and charging for illegal and poorly planned introductions of exotics.
- Enact national minimum standards for state fish and wildlife laws and provide incentives for and sanctions against states to encourage them to do so; a reliable source of funds would be needed.
- Strengthen, redefine, and update federal laws restricting the entry of noxious weeds and seeds and beef up enforcement. The government took eight years to place 93 species on the current list of noxious weeds, yet at least 750 weeds meeting the legal definition of "noxious" remain unlisted.
- Clarify the federal role in regulating the transport of weeds between states and make planting, distributing, and possessing noxious weeds with intent to distribute them illegal under almost all circumstances.
- Spend more money on programs to control noxious weeds, including those on federal lands.
- Require anyone introducing foreign plant material to conduct scientific evaluations for its potential for spread and damage to native species.
- Ensure that decision-making processes regarding exotics are documented, clear, open to public scrutiny, and periodically evaluated.
- Assign broad and explicit responsibility for the control of exotics that damage natural areas to a federal agency and provide money for its implementation.
- Allocate more money, or charge national park entrance or user fees, to perform large-scale control and eradication measures against exotics.
- Impose new restrictions on federal agencies and others that use federal funds to introduce non-native fish and wildlife.
- Set deadlines for new, comprehensive regulations on the importation of unprocessed wood.
- Establish a surcharge on boat and boat trailer licenses to fund control programs.

Source: U.S. Congress, Harmful Non-Indigenous Species, 15, 16, 18, 20, 26.

You can gauge how threatening an exotic insect is by just counting the number of tree species it savors. You can also estimate its rate of spread. The European gypsy moth has dispersed quickly in the United States by laying egg masses on camper vans and boats and hitching a ride down the highway, to the Blue Ridge among other places. Without that human help it travels relatively slowly—only a few hundred yards a day—because the female does not fly. But Asian gypsy moth females can fly, reportedly up to 60 miles in a single day through the forests of Europe and Asia.

William Wallner, a Forest Service research entomologist, has watched these Asian moths in his laboratory's "flightmill," a kind of treadmill for insects, as they fan their wings for about four minutes to warm up. Then they take flight—some, for several hours.

July 4, 1993, was about what you'd expect at the port of Wilmington, North Carolina. Hot and humid, and another military cargo ship, the *Advantage*, was just arriving. It docked at the Sunny Point military shipping terminal, completing the ten-day transit over the Atlantic from Nordenham, Germany. Hundreds of World War II–era containers that resembled big metal boxes were stowed in its four big holds and stacked on deck. They had been in storage, outdoors under some trees, probably for years.

Philip Bell, a federal plant protection officer, boarded to look over the ship's manifest, examine the garbage, and "take a gander around." Because it was a holiday, nothing much moved, but by the time he arrived the next morning around 9, the crew had begun off-loading. "When I got there," Bell recalls, "I went and saw the captain. We discussed where the ship originated. A hundred sixty containers had come off by then. All of a sudden I saw some flying moths. I grabbed one and looked at it, and I saw it was a gypsy moth. We stopped the off-loading. There was thousands of them flying around. I talked to one of the stevedores that worked the ship. He had noticed that when they opened up the hull that morning, some started flying out. The day warmed up and they just exploded."

Bell was quickly on the phone with his employer, the Animal and Plant Health Inspection Service of the Department of Agriculture—APHIS—to try to identify the insect. "They asked me if the females could fly," he recalled later. "I went back down to the ship and grabbed one of the females and flipped her, and she flew. I went right back to the phone." They were certain then that it was an Asian gypsy moth. DNA analysis later confirmed it.

"Any time you have a foreign pest flying around on the coastline, there's urgency," Bell says. "We had a sort of a little power struggle there to get the ship back out to sea until all lifestages of the moth were dead." By the time the containers were reloaded, it was well past high tide, and the ship couldn't leave. Toward midnight, it finally cleared port and moved out to sea 5 miles beyond Bald Head Island, the farthest point of land. "We felt that was safe," Bell says.

The Sunny Point area was quickly saturated with hundreds of traps that use ultraviolet light and moth sex hormones as lures. The ship underwent the largest fumigation in the history of APHIS. But those were only the initial moves in what eventually became a three-year, $3 million effort to try to make sure no moth survived. Pesticides were sprayed widely enough that local schoolchildren remarked the absence of butterflies the following spring. Thirty thousand traps were placed in North and South Carolina.

Eleven Asian moths were caught in 1994; some had moved inland as far as 10 miles. The following year, three were caught and then none. Further monitoring was planned for a few sites, just to be sure.

It was a success story, so far as is known, if rounding up genies after they are out of the bottle can be thought of as "success." Asian gypsy moths have also been found three times on ships entering Pacific Northwest ports. The species "has the potential to be the most serious exotic insect ever to enter the U.S.," one agriculture official has warned.

"With as much movement of people and commodities as there are," Bell says, "I wouldn't be surprised to see them again. With the budget cuts coming along now, it's getting sort of tough. It just scares me to think what's going to happen down the road."

William Wallner checked back to see what became of other shipments of military equipment arriving in Wilmington. They were dispersed to forty-eight different locations throughout the United States. "There are no guarantees that these were either free from, or infested by, Asian gypsy moth," he says.

The Forest Service's Robert Anderson, a twenty-five-year veteran of such battles, recalls the demise of chestnut, of dogwood, of oak, and other species and peers ahead: "What we're going to see in the next 40–50 years is an ecosystem that's never been seen before. There will be a shift in animal populations, too. A hundred years after that, it'll turn again, depending on what exotic pests we introduce. We're moving to a red maple–black gum–tulip poplar forest, according to our surveys. . . . If we introduce an exotic pest that likes tulip poplar or red maple . . ."

Maple trees are cherished by the Puerto Rican and Polish Americans who populate the Greenpoint section of Brooklyn. One of them noticed, on August 19, 1996, that sugar maples on his block seemed to have been vandalized, as if someone had drilled holes in them, and he reported it to the Parks and Recreation Department.

"So the following day I went out," forestry inspector Harry Rothar recalls, "and all of a sudden I look up and I see this big, beautiful black beetle with white spots all over it. I had never seen anything like that before, so I collected one, and I brought it back to the office and I showed it around. No one knew what it was. . . . I spoke to numerous people up in the area and they had seen it for three or four years, brushed it off their trees and when it landed on the ground, stepped on it." The beetle may, in fact, have appeared in Brooklyn as long ago as the late 1980s.

Entomologists soon identified it. More than an inch long, it carries even longer, gracefully curving antennae banded in black and white. Its glossy, coal-black body is covered with irregular bright white spots. The Chinese call it the "starry sky beetle," and it is considered a major pest throughout its range in China, Japan, and Korea. In North America, that range corresponds to all of the area—including the Blue Ridge—between the latitudes of Milwaukee and Cancun.

The starry sky beetle is a voracious feeder in all maples, including the red maple and poplar that are increasingly common in the Blue Ridge. But the list of its known "host" trees—there may well be others—includes dozens of wild and domesticated species: willow, chinaberry, mulberry, plum, pear, black locust, elm, and horse chestnut among them.

"Brooklyn is a major shipping port and we suspect that the starry sky beetle entered New York on wooden dunnage [log braces used to support cargo in ships] or crating material," writes Forest Service research entomologist Robert Haack, whose reconstruction of events is the basis of much of this account.

Adult beetles may fly distances of a thousand yards in search of new trees to infest. The females chew through bark, then turn and lay one egg in the pit they have made. They lay twenty-five or more eggs during their several-week life span. Eggs hatch after a week or two. Larvae bore in and create a succession of galleries, feeding on cambium, then sapwood, then heartwood, pushing wood fiber and feces out behind them as they go.

This beetle may be a "slow spreader." On its own, it might take decades to centuries to get from Brooklyn, say, to the Blue Ridge. But APHIS administrator Victor Mastro says the more realistic figure is about twenty-five years, because traveling humans would carry them south. The beetle can hitchhike easily in firewood, for instance, which is often shipped long distances.

As if to illustrate that possibility, a second beetle infestation was also found on Long Island. It probably occurred when tree trimmings from Brooklyn were carted to a wood-chipping plant in Amityville. And one truckload of firewood—it turned out to be clean—was intercepted while en route from Long Island to Montreal.

A five-year state and federal quarantine was thrown up around the infested area to try to prevent the movement of tree trimmings and wood. Plans were laid for years of surveillance across the river in Manhattan and in other nearby areas. Infested trees were cut down.

"As you can imagine," Haack said while the job was under way, "there's a lot of politics involved. Who's going to pay for it? The state wants the federal government to pay, the feds want the state to cost-share the program . . . the sugar maple industry and the whole tourist industry in New England are quite worried that this thing is going to get out and devastate the maples. So there's a lot of people putting pressure. They want this insect eradicated.

"And then there's all the local pressure. There's certain streets where almost every tree is infested, so you go from a street that is lined with large maple trees to cutting them all down. A lot of people have a nice tree in their front yard and then here comes somebody with a chain saw and cuts your tree down. And it's expensive."

More than a thousand infested trees were found in Greenpoint and Amityville. After they were felled, the stumps were ground down 6 to 18 inches below street level. The trees were fed to a chipper, and the chips were incinerated.

Exotics are one of those "ounce-of-prevention" kinds of problems. More vigilant inspection requirements for international trade would have prevented this and other infestations before they ballooned into a severe threat. The "pounds of cure" are hugely expensive and only occasionally effective. Despite the thoroughness of the efforts to eradicate the starry sky beetle, federal officials say the results are not certain and that there may be other "hotspots" outside Amityville, still to be discovered.

The beetle has also been intercepted in Loudenville, Ohio, and in Vancouver, British Columbia. Wood bearing evidence of infestation has shown up in crates of steel water pipes, granite blocks, limestone blocks, porcelain tiles, and many other products shipped in from China.

"Once it gets to the U.S., it's not like it sits in a port area," Mastro says. "Wherever that crating material could end up, that's where those beetles could be."

"If you think about this thing being loose in a forested area, where you have no idea where it is . . ." Mastro adds, his voice trailing off. "And that may be the situation right now. I don't want to hype this, either. My gut tells me we have not found all the infestations that are out there."

The authors of a recent research report on exotics departed from the dry and measured language of entomology to write: "When the outrageous economic and ecological costs of the wanton spread of existing exotics and

SOLUTIONS 4

Biopollution: Whose Problem, Who Pays?

Research entomologist William Wallner is a consultant to officials in the United States, Russia, Mexico, and New Zealand, among others, who are trying to slow the movement of invasive forest pests. Some of his recommendations:

- It is much better to stop exotics that "stow away" on shipping with thorough inspections in the countries where they originate than to try to detect them when they arrive here. This also minimizes inspection delays, the need for fumigation or other treatment, and confiscation of products in our ports.
- Countries lacking expertise or resources might be subsidized by wealthier countries to take preventive steps—much cheaper than trying to solve the problem at the "receiving end."
- International cooperation between scientists is a critical need. The United States should support an immediate global scientific effort to compile and categorize lists of invasive pests for sharply increased quarantine and surveillance.
- The ecological and economic costs of pest invasions should be considered on a global basis, among nations and trading blocs, and they should immediately standardize procedures and identify the highest-risk pests.

continued entry of new ones become common knowledge, it is inevitable that there will be a public outcry for action to mitigate the potentially dire consequences."

None too soon. Doing without more effective protection along these distant borders of the Blue Ridge is like flicking lighted cigarette butts into dry tinder and—despite evidence of a repeatedly charred landscape—hoping that nothing much will happen. For now, the pace of new introductions of exotics is undeniable evidence that our defenses are feeble. Pine shoot beetles, already introduced to the Great Lakes region a few years ago, have shown up as far south as Maryland despite a federal quarantine. European spruce bark beetles and red-haired pine bark beetles have made it to shore. They have been found living in piles of shipping refuse up and down the east coast and eradicated.

"Things get by you. We don't have security at all the piers in the country," one APHIS entomologist says. "Customs tries to handle that, and we try to handle our part of it, but it just doesn't happen."

Reports of ships that arrive in the United States with tree-killing pests

- DNA tests should be used to identify high-risk pests now, in each country where they originate. Then regulatory officials would know exactly where a new invasion has come from, and they can track down how it arrived.
- Alert import-export businesses as to pathways that pests use to enter North America and share the information with trading partners.
- Make the user fees already charged for inspections of exports and imports available to APHIS to fund research on identifying, detecting, and eradicating exotic pests.
- Enact laws that will permit U.S. government agencies to sue those responsible for all losses associated with intentional or accidental introductions of exotic pests. Penalties now are minuscule.
- Begin immediate research on identifying and analyzing the behavior of exotics already here to assess how great a risk/threat they pose.
- The United States should reduce its dependence on imported raw wood because of its potential for the introduction of new exotics that attack forests.

Sources: Wallner interview and Wallner, "Invasive Pests," 167–80.

and have to be "sanitized" are a daily occurrence, according to APHIS personnel. The agency's efforts, short handed, hamstrung by weak and confused laws, and subject to political pressure, were criticized in a recent book-length federal analysis of the threat of exotics. Few of the report's many recommendations have been implemented.

Without quick and emphatic changes in policy the Blue Ridge faces a continuing influx of trouble. "Fifty years from now we certainly will have new exotics and as a result of that, forest composition will probably be less diverse," Haack says. "The woods will still be green, but the species composition will be more homogeneous. . . . As a result of all the defoliation of oak by the gypsy moth the forest seems to be shifting to red maple. This insect in New York prefers maple. The beech bark disease prefers beech. I'm not sure what's going to be left to take their place."

The toll among Blue Ridge trees mounts. Along with the cases already cited, the butternut is being decimated, and even faces near-extinction in the wild, as a direct result of biological pollution (see Solutions 5). Oak,

SOLUTIONS 5

Breeding Away from Exotic Pests

On two small tree farms near Abingdon, Virginia, researchers have been working for a decade to resurrect the American chestnut, which essentially disappeared from the Blue Ridge—and everywhere else in eastern North America—six decades ago.

When chestnut blight struck, the dream of finding a resistant variety in the wild gradually faded as the disease marched on through the mountain forests unimpeded. Breeding experiments using the blight-resistant Chinese chestnut, a small orchard tree that bears little resemblance to the majesty of the mature American species, were also disappointing.

The American Chestnut Foundation's project in Abingdon is different, however. It is based on "backcrossing" successive generations of American and Chinese hybrids until a nearly pure American form, but with gene-based Chinese resistance transferred in through breeding, results. Sometime after the turn of the century a blight-resistant variety will emerge if all goes according to plan.

The fate of butternut trees, whose commercial value is very high because of their use for woodworking and carving, may also hinge on resistance to an introduced fungus.

"Butternut continues to decline," says Robert Anderson of the Forest Service. "As far as we can tell the whole range of butternut is now infected. The disease has extended up into Canada. Throughout the Blue Ridge area and every place we've looked we have found infected trees. Our suspicion is there's total infection throughout the range.

Eastern hemlock, ash, beech, maple, and pine will, barring a miracle, be infested by still other severely destructive exotics that are already established and whose ranges, according to federal officials, are spreading south.

Diseases and insect pests of trees are only one battalion among the exotics marching on the Blue Ridge ecosystem, though. "A major problem is noxious weeds," says the Forest Service's Robert Anderson. "We're getting incredible numbers of introductions of weeds from other countries that are highly competitive with our native vegetation. Vines like oriental bittersweet, exotic grasses, mile-a-minute weed—it grows like crazy, has thorns on it, nothing will eat it—there's a tremendous number of exotic plants that are being introduced or are already here."

Purple loosestrife, which has been called an "ecological disaster," is increasingly common in the national forests and on the verge of invading Great Smoky Mountains National Park. In some landscapes outside the Blue Ridge, it has physically displaced half of the native plant life, by

"In seventy-five to one hundred years, the species may potentially be lost. The disease attacks everything from mature trees to poles to small trees to nuts. The fungus is systematically eliminating the tree."

But that is not quite the end of the story, Anderson says. Two coves of butternut trees have been discovererd in the Pisgah National Forest in the Blue Ridge, in which about a third of the trees are genetically resistant and disease free, though neighboring trees are dying, blighted by hundreds of fungus cankers. A few resistant trees have also been found in Virginia, Arkansas, and Kentucky. "We are very encouraged, especially with the trees in the Pisgah," Anderson says. They have been turned into a seed production area, which is carefully thinned, fertilized, and tested.

Forest Service personnel are also on the lookout for hemlock trees that show resistance to the hemlock woolly adelgid.

An intensive search has located, in Catoctin Mountain Park, Maryland, a few dogwoods that are somewhat resistant to the exotic dogwood anthracnose virus. One tree, Anderson says, is very resistant: "No matter what we throw at it, it walks away." Now residing in a greenhouse at the University of Tennessee, it has been "multiplied" several times, and its resistance factor is inheritable. Perhaps it will be the basis for some future attempt to restore a fraction of the lost dogwoods to the forest.

Source: Robert Anderson interview.

weight. Much of that, in turn, was an important source of food for other species. Purple loosestrife has also contributed to the decline of birds and turtles in such places by destroying their habitat.

Of the total of 1,364 plants known to occur in Shenandoah National Park, 224—about 16 percent—are exotics. Here are forest ecologist James Åkerson's notes on the campaign there:

Able to get rid of:
1—kudzu; its area is still small.
2—perhaps Japanese knotweed; this will be a hard battle in several concentrated areas; it too is not yet widespread.

Contain them:
1—Asian bittersweet
2—Japanese honeysuckle
By this, I am quite broad in interpreting "contain." It will be an ongoing battle.

SOLUTIONS 6

Biocontrol

Something strange happened in the northern Blue Ridge in the late 1990s. For the first time in several years, the great wave of gypsy moth infestation that had been defoliating and killing thousands of trees, especially oaks, had subsided almost to nothing. The remarkable absence was welcome but probably temporary.

The advance had been stopped by a fungus purposely imported to North America to kill gypsy moths, an example of a strategy for fighting exotics called "biological control." The spore lands on the hairs of the moth, cuts through its skin, and reproduces inside its victim.

Experts are not sure what the trajectory of gypsy moth populations will be in the future, but, alas, none are predicting its disappearance from the Blue Ridge. Still, biological control holds promise for at least slowing the spread of such pests, and perhaps, in some cases, containing or even eliminating their populations.

Sometimes the plan is to employ a fungus, bacterium, or virus to control an insect; sometimes it's to get one insect to attack another; sometimes it's finding a virus to control a fungus.

Biological control is a patient and uncertain business. It often disappoints when introduced parasites and predators fail to reproduce or to perform as hoped. And some of the organisms introduced as controls in the past have proved to be more harmful than the exotics they were supposed to help get rid of. Years of testing are now required to try to make sure any newly introduced biological control agents pose little risk to other species.

Other examples of "biocontrol" campaigns in the battle against exotics in the Blue Ridge:

- Research entomologist Mark McClure has been studying a small beetle he discovered preying on the hemlock woolly adelgid in Japan. After five years of study, 2,430 adult beetles were set free in a Virginia hemlock

Let 'er rip:

All the rest will have their way because of our inability to eradicate or control them. They will become a solid part of the Blue Ridge environment. Perhaps they already are and we haven't admitted it. . . . They include:

Tree of heaven, garlic mustard, Japanese grass, multiflora rose, Chinese yam, and all current introduced insects and diseases.

. . . As I and we learn more about other exotics here, we may find they too can be contained either because they aren't yet widespread or because we possess the technology and will.

forest not far from the Blue Ridge Parkway in 1997. After two years, McClure should be able to tell if limited releases are working.

- At about the same time, Forest Service entomologist Michael Montgomery and exotics specialist Kristine Johnson of the National Park Service were in the Yunling Mountains, in the foothills of the Himalaya in China to study a different but related beetle that also feeds on the hemlock adelgid. The climate there, where hemlocks grow at 7,000–10,000 feet, is somewhat like that of the Blue Ridge. It was the latest of Montgomery's several trips. "These seem to be specialists on adelgids," he says, "and that's what we want." Similar research is under way to find organisms that prey on the balsam woolly adelgid.
- A virus that attacks and weakens the chestnut blight fungus is the focus of an experiment planned for Shenandoah National Park. Plant pathologist William MacDonald of West Virginia University is applying the virus to blight cankers in a grove of chestnut sprouts. Several experiments have prolonged the survival of the trees by weakening the fungus in this way. The trick is to induce the virus to spread naturally.

It is tempting to hope for breakthroughs in fighting exotics, but researchers say the reality is that years of patient research and consistent funding are required for each such experiment. Biological controls, when they are successful, still must be used along with other eradication and control techniques—especially stopping more new introductions of exotic pests. "No 'silver bullets' exist for . . . control," a federal study of the exotics problem warns.

Sources: Hajek, Humber, and Elkinton, "Mysterious Origin," 31–42; U.S. Congress, *Harmful Non-Indigenous Species,* 151; McClure interview; Montgomery interview; MacDonald interview.

Great Smoky Mountains National Park is now home to 320 or so exotic plants—more than a fifth of the total number of plant species in the park—including two dozen that are considered "invasive and aggressive, capable of displacing native species, and noxious to native plant communities."

As at Shenandoah, some invaders such as kudzu are controllable, given enough time and money. Others, like Japanese grass, are already so well established in the park that the best-case scenario is just to limit their spread.

"I'm getting more and more worried about the Chinese yam," the park's plant pest specialist, Kristine Johnson, says, though she also has a wary eye

SOLUTIONS 7

Garden-Variety Threats

One source of exotic plant invasions is gardens and landscaping. Botanists recommend that only native plants be used, because garden exotics can "escape" and spread with ease into protected natural areas. If exotics are used, they should be certified as "noninvasive" and free of any risk of introducing exotic insects or of hybridizing with native species. "The Blue Ridge has a spectacular flora. We need to work with that flora," says University of North Carolina botanist Peter S. White, who is also a director of the North Carolina Botanical Garden.

Source: Peter S. White correspondence.

out for purple loosestrife and mile-a-minute and occasionally has dreams about the park's persistent kudzu.

The yam is like a series of tiny potatoes that sprout prolifically and wind their tendrils around other plants. It can be carried by animals and water to new places, has spread throughout east Tennessee, and is already in the park.

Herbicides can't be used against the yam, because it has the same growing season as much of the native vegetation. "It's not on any of our big target lists," Johnson says, "but it's something we're seeing more and more. It's showing up in my sister's garden, my mother's yard, at my boyfriend's. Everyone's saying, what is this stuff?"

The federal report on exotics warns, "Concerns are increasing that the ecological changes overtaking the parks may be so severe that they will eliminate the very characteristics for which the parks were originally established. Unchecked, such changes eventually will eliminate the national parks' role as a caretaker of U.S. ecosystems and indigenous species."

Wild hogs exemplify the behavior of many exotic pests. Descended from imported European animals that escaped from hunting preserves in 1912, they now eat, uproot, or trample at least fifty species of plants in Great Smoky Mountains National Park, and they can reduce the plant cover on the forest floor by 95 percent. Their rooting displaces animals like voles and shrews and causes soil erosion that fouls streams. They eat small animals, including rare salamanders and snails, and compete with native species for food. And they have proven to be hard to kill off, partly because they've won a constituency among local hunters, who favor the hog's presence here as a game species. Hunters have sometimes felt that the hog has been discriminated against, as a newcomer.

All this talk of excluding exotics, sometimes called "alien" species, may seem to contradict concern about preserving biodiversity, if we think of diversity as "the more species the better." Some observers even sense, in efforts to fend off exotic species, the phobic side of our national debate about immigration policies for humans.

But whatever the merits of immigration policies, comparison with exotic plants and animals is mistaken. The problem with exotics is not their foreignness but the fact that they go "hog wild"—they tend to overwhelm and kill off native plants and animals, altering the whole ecosystem. In other words, the kind of diversity that exotics bring is brief and unsustainable. It leads eventually to fewer species, to biological "sameness"—a kudzu blanket of homogeneity that is the opposite of biological diversity. Exotics are a factor in more than 25 percent of the cases in which endangered species are declining and a major cause—or the major cause—in about 8 percent of them.

One barrier to a clearer public sense of the magnitude of the exotics problem is what might be called its story line. "Here come the alien bugs" sounds like a too-familiar horror movie—like *Jaws*, actually. That movie's subtheme, it may be recalled, was that our own species seems at times too distracted or short-term-oriented to organize an effective resistance.

In 1987, a north Georgia aquaculturist named David Cochran began importing live white sturgeon, a few days old, from California to his Blue Ridge Mountain Fisheries. There at Wilson Spring, after ten or twelve years, they would mature to yield a lucrative and recurring harvest of caviar, via cesarean section. By their second year, they might grow to 15 or 20 inches. White sturgeon don't seem much like "Jaws," perhaps. Native to the Pacific states, they have small mouths and no teeth.

In April 1989, a puzzled David Whitmire caught what he thought was some kind of freshwater shark on the Coosa River in Alabama about 150 miles downstream, as it happens, from Cochran's hatchery. It was later identified as a white sturgeon, and it measured out at 20 inches. Biologists figured it was discarded from someone's aquarium, and they may have been right. The species is legally sold in pet stores in Georgia and Alabama, and aquarium owners sometimes dump their erstwhile pets.

In the winter of 1990, a flood swept through the area, inundated Cochran's tanks and ponds, and washed hundreds of his white sturgeon out into the woods to die. Others, however, may have gotten into adjacent Kenyon Creek, which feeds into the Coahulla, which is, in turn, a tributary of the Conasauga. And the Conasauga is one of the state's last undammed, free-flowing rivers. Upstream, it leads into the Blue Ridge and Tennessee, looping back into Georgia as it draws from a web of cold, rocky mountain creeks that closely match the white sturgeon's native spawning grounds.

Downstream, the Conasauga joins the Coosawattee to form the Oostanaula, which joins the Etowah to become the Coosa, which merges with other rivers and eventually flows on out to the Gulf. "Whether any of them actually were washed downstream we don't know," Cochran says, "because there were measures taken to keep them confined . . . but I can't say absolutely that none of them escaped."

A few months after the flood, the Georgia Department of Natural Resources raided Cochran's business and confiscated his 1,223 sturgeon. He had waged a running battle with the agency for several years about what he regarded as burdensome and unnecessary regulation of the fish-farming industry. In launching the white sturgeon venture, Cochran had not obtained a special permit that the state said was required for importing exotic fish.

He claimed that he had not needed a permit, because he interpreted "exotic" to mean "not native to the United States." The state argued that the term was as plainly understood as the reasons why exotics are regulated: they can devastate native species. Cochran won his case, however, and later won a $775,000 settlement from the state to pay for damages to his business, bankrupt after the raid. The statute has now been changed to define "exotic" fish as "not native to Georgia."

In November 1992, still another fisherman, Milton Truss, caught a 2-foot-long white sturgeon in a reservoir along the Coosa River. Alabama fish and game officials couldn't fathom the presence of a Pacific fish in Deep South waters.

A free-living population of white sturgeon somewhere in the Southeast is a remote scenario, officials acknowledge. The larger point is that there are now many paths by which destructive exotics can enter the Blue Ridge ecosystem. And few roadblocks, only lightly patrolled.

In any case, who knows? "Maybe in a few more years, if suddenly there's lots of little white sturgeon showing up, it'll mean that a couple of them managed to get together and do it," Georgia fisheries biologist Michael Spencer says. "I hope that won't happen."

White sturgeon can live for eighty years. They would certainly make a fine catch. At full growth—up to 20 feet long, and sometimes more than 1,800 pounds—they are the largest freshwater fish in North America.

"My immediate interest is this," says University of Georgia biologist Byron Freeman: "What is the potential impact of a large, fish-eating, mollusk-eating predator in an aquatic ecosystem that is full of teeny species of mussels and fishes? This is the last hurrah for a lot of these rare animals. It's a real bad potential problem here." White sturgeon also eat fish

eggs, and they can carry an assortment of viral diseases such as herpes—
strains that may be new to the fish of the Southeast.

After their release from state custody, Cochran's remaining 850 or so sturgeon were sold by the bankruptcy trustees. They were moved, to inhabit a private lake near Rising Fawn, Georgia, but just over the border in northeastern Alabama. In that state, exotics are even more lightly regulated. The lake drains into a local stream, and it runs—via Lookout Creek and the Tennessee River—back into Georgia.

L ike startled shadows in a pool on the Saint Mary's River, brook trout flick and glide. Slime-slick, dark olive when seen from above, bearing a calligraphy of speckles, they evolved as a separate species perhaps a million years ago.

Throughout the Blue Ridge, these natives have been driven from lower elevations by exotic brown and rainbow trout and by water that is now too warm. Trees next to trout streams have been cut for lumber, farming, or urbanization, and their cooling shade is gone. In what is now Great Smoky Mountains National Park, brook trout habitat has diminished by 70 percent since the turn of the century.

High, chill, isolated streams like the Saint Mary's, which tumbles for a mere 7 miles down a narrow Virginia valley, are a last refuge. And a brook trout's life, it turns out, is linked by air and water to the dirt, the trees, and the clouds overhead.

If you're looking out over the riffles, and the sun they reflect, idle questions may come to mind. Where does this water come from? Which skies and soils does it move through, as it finds its way here and then on south and east to the Chesapeake Bay? Scientists have been asking the same questions, but with increasing urgency in recent years, as life ebbs from mountain streams like this one. Their research traces, on continental maps or through inches of soil, the paths the water follows, and they analyze its shifting chemistry.

It's a complex story, and much of it is still incompletely known, but a few central themes have emerged, and the Saint Mary's displays them with unwelcome clarity. Seven of the twelve fish species there, including the exotic rainbow and brown trout, as well as native chubs, suckers, and dace, have vanished during the past two decades along with the mayflies they fed on. Of the five remaining fish species, four are essentially gone. They are detectable only at the last and farthest downstream of several monitoring stations. Seven out of seventeen kinds of insects have disappeared, and others have declined sharply.

The brook trout remain—if they remain—because they have withstood the increasing acidification of the water and surrounding soil caused by regional air pollution that yields acid rain.

Virginia state fisheries biologist Larry Mohn worries that, after a stay of seventy-five thousand to one hundred thousand years, even the brook trout may succumb to acid rain. In at least two of the final five years of the century, scientists could find no trace of a spring hatch of brook trout in the Saint Mary's.

Mohn has seen the decline of this watershed as it is charted in biological survey data collected since the 1930s. He has also seen it at first hand. "I used to fish a lot, and I'd take friends in there fishing," he says. "I look at it now and see a fairly sterile stream. Its bottom is extremely clean. You don't see any aquatic vegetation or any vegetation attached to the rocks. You can just tell it's not what it used to be."

Acidification brings the arrival of a few new acid-tolerant species of insects such as black flies and some plants. But there are fewer species overall—in other words, a general narrowing of biological diversity.

The Saint Mary's drains a 10,500-acre watershed that is legally designated as one of the nation's wilderness areas, in a section of the George Washington National Forest. On the Mine Bank Trail, which descends from the Blue Ridge Parkway down toward the river, droning engines and keening tires are soon beyond earshot. Even on a fall weekend, when the slopes are streaked with tangerine and crimson and parkway traffic peaks, only a few hikers ford the river where the trail crosses.

The Wilderness Act of 1964 defines these as places "where the earth and its community of life are untrammeled by man, where man himself is a visitor who does not remain . . . an area of undeveloped Federal land retaining its primeval character and influence . . . which is protected and managed so as to preserve its natural conditions."

Unfortunately, federal officials find this responsibility impossible to meet. The Blue Ridge ecosystem includes twenty-nine federally protected wilderness areas like the Saint Mary's (fig. 4). It also has some of the most acidic rainfall in the United States. Like the rest of us, the Blue Ridge and its wildernesses breathe a non-"primeval" atmosphere. The sky overhead is shared with Cleveland, Ohio; Knoxville, Tennessee; and Charleston, West Virginia, among many others. "Preserving natural conditions" in the wilderness areas is like putting a baby in a room full of smokers, then admonishing the baby-sitter to make sure the child breathes only the clean part of the air.

Weather usually arrives in the mountains from the west and south, where it picks up an assortment of cast-off and highly reactive chemistry— the exhalations of industrial activity and auto traffic down below (figs. 5, 6). Anywhere along this path, hundreds of miles long, wind and water vapor may interact with plumes of pollution made up of oxides of nitrogen and sulfur. They become nitrate, sulfate, nitric acid, sulfuric acid, and other compounds.

Most of the sulfur and nitrogen oxides are pumped into the atmosphere by automobiles and electric power generating plants that burn coal. The power plants often work at full capacity, generating peak pollution, during the summer when extra electricity is needed to run millions of air conditioners. Only about 6 percent of the sulfur and 15 percent of the nitrogen

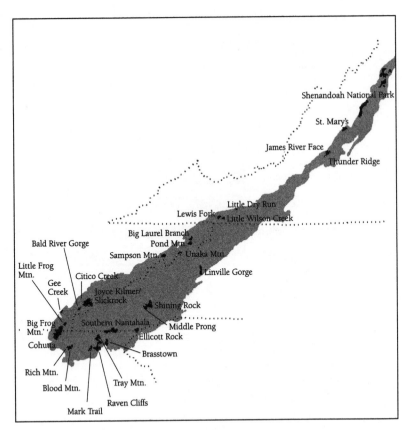

Figure 4. *Wilderness areas in the Blue Ridge.* Most of these legally designated wilderness areas are within national forests. They are managed as protected areas where natural ecological processes operate freely and human influences are minimized. (*Sources:* Keys et al., *Ecological Units of the Eastern United States,* CD-ROM; Hermann, *Southern Appalachian Assessment GIS Data Base,* CD-ROM)

oxides emitted into the atmosphere in the United States come from natural sources. Humans contribute the rest.

These chemicals interact with water vapor and sunlight and waft along with the prevailing winds. They "rain" on the Blue Ridge in several forms: as rain, snow, fog, gases, or dry particles. From there, water moves the chemicals along to other interim destinations in the ecosystem. Science refers to these journeys through soils, streams, and living organisms as "chemical fates."

To explain what happens when they descend to earth we need some grade-school chemistry. The strength of acid and its chemical opposite, alkalinity, is measured on the pH scale, which runs from 0, the strongest acid, through 7—neutral—to 14, the most alkaline.

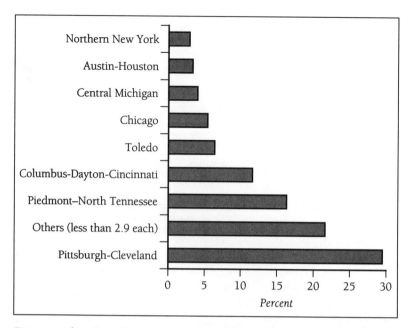

Figure 5. *Where does sulfur-based air pollution at Shenandoah National Park come from?* These percentages are averages from 1983 to 1987—the most recent data of this type available. Though they appear to be precise, they are not. Rather, they give a general indication of the widespread and distant sources of this type of air pollution. (*Source:* Gebhart and Malm, "Source Apportionment," 907)

So as pH falls, acidity rises. Distilled water has a neutral pH of 7; lemon juice, 4; vinegar, 3; stomach acid, 2; sulfuric acid, 1. Nonpolluted, natural rainfall is slightly acid, with a pH from 5 to 5.6, depending on the locale. In the Blue Ridge, rain pH averages 4.3 to 4.5 and is, at times, far more strongly acidic than that. The pH of high-elevation fogs has been measured at well below 3.

When we hear such numbers, we usually think in terms of rulers, or measuring cups, or bathroom scales. A rise or fall measured in tenths hardly seems alarming. But the pH scale is different. Each full unit of decrease in pH (from 6 to 5, for example) represents a tenfold increase—that is, a 1,000 percent increase—in acidity. Similarly, a decline of 2 on the pH scale represents a one-hundred-fold, or 10,000 percent, increase in acidity.

The average pH of the Saint Mary's River has declined during the past sixty years from 6.8 to about 5, or nearly 10,000 percent. In this as well as in other, less acidified streams in the region, pH can drop further, as low as 4.7 during storm or snowmelt runoffs. These pulses of highly acidic water flush both soil and aquatic life with acutely toxic, sometimes lethal chemistry.

Acid's effect on the ecosystem when it reaches the ground depends to a

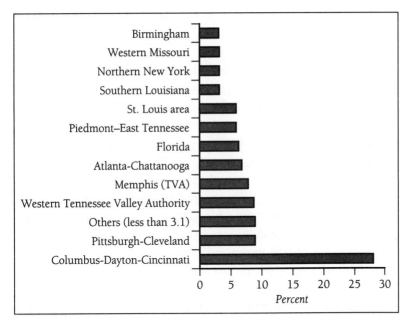

Figure 6. *Where does sulfur-based air pollution at Great Smoky Mountains National Park come from?* These percentages are averages from 1983 to 1987— the most recent data of this type available. Though they appear to be precise, they are not. Rather, they give a general indication of the widespread and distant sources of this type of air pollution. (*Source*: Gebhart and Malm, "Source Apportionment," 908)

great extent on local geology. A watershed underlain with rock that has plenty of alkalinity in its chemistry is more effective in neutralizing acid. So is rock that weathers more rapidly, breaking down into acid-neutralizing molecules. The highly alkaline limestone geology of the Shenandoah Valley, for instance, protects its soils and streams from the effects of acid rain. But the rock under the Saint Mary's watershed, and much of the rest of the Blue Ridge, does not.

Think of soil's ability to neutralize acid as something like a big sponge on your kitchen counter. As you pour a pitcher of water onto the sponge, some of the water is absorbed and some runs off. As the water continues to pour, though, the sponge's capacity diminishes, and more of the water runs off. Eventually, the sponge is saturated, and none of the incoming water is retained.

In the same way, as acid arrives on mountain soils, some of it is neutralized or "buffered" by their alkalinity, and some is not, so it runs on into streams. At current rates throughout the Blue Ridge ecosystem, incoming acid far exceeds the slow addition of alkalinity from mineral weathering. As a result, the buffering capacity of the soils is declining. The organisms whose health or survival depends on this buffer are living, then, on borrowed time. As it is exhausted—a matter of a few decades for many watersheds and sooner for others—the "sponge" will cease to function. Nearly all of the acid that reaches the ground will run off into the streams, almost none of it will be neutralized, and stream acidity will rise sharply. "The result of these changes will be catastrophic to ecological balances in these streams," the National Park Service has predicted.

Vulnerability to acid rain is the rule rather than the exception (plate 5). North of Roanoke, about four out of every five Blue Ridge brook trout streams are "sensitive" or "extremely sensitive" to acidification. Farther south, the mountains have deeper, more forgiving soils, with the notable exception of the high peaks and ridges. There, the fog and rain are harshly acidic, and the soils are thin.

As landscapes acidify, brown and rainbow trout disappear at earlier stages, but brook trout have no love for highly acid streams either. When offered a choice in lab tanks, they will consistently seek out and swim into less-acid water. They do best, growing faster and living longer, in water with a pH of 7 or 7.5. "Sublethal" symptoms such as weight loss and diminished reproductive success begin to appear at pH levels below 6.5.

Except at extreme levels that cause "acid shock," the fish only rarely succumb to acid itself. Instead, they fail to reproduce successfully, or they die from a side effect: aluminum poisoning.

This happens because as water moves acidity through the ecosystem, it does not leave soil chemistry unchanged. It dissolves some elements, such as aluminum, which is the most abundant metal in the earth's crust. Like many dissolved metals, aluminum can be lethal at high concentrations when it makes its way into the organs of living things.

Trout bury their eggs in stream gravels in the fall, and the eggs hatch during spring runoff, just when the heaviest concentrations of acid and mobilized aluminum arrive. At a pH of 5, adults will survive, but newly hatched brook trout begin to die. At 4.5, all ages die.

University of Virginia fish physiologist Arthur Bulger, who autopsies such cases, explains that the "site of toxic action" in fish is the gills. Gills draw oxygen from the water, and they also maintain the proper mix of dissolved minerals—electrolytes—in the fish's blood. "In fact, the mix of minerals is very similar to that of humans," he says. Humans and freshwater fish both get needed minerals from their food. But the fish can also absorb other minerals, via their gills. ("Humans can't do that," Bulger notes.)

Aluminum, however, cripples the enzyme process that allows the fish to pull in electrolytes from the water. Aluminum also loosens the connections between cells in the gill membrane, allowing electrolytes concentrated in their bodies to leak out of the fish. Their red blood cells swell, so the blood gets thicker and much harder to pump. This, in turn, disrupts the proper functioning of nerves and muscles.

"Poisoning" is, Bulger points out, a technically accurate term for this sort of death. Sometimes it is cumulative and chronic, as when fish populations, and the insects they feed on, sicken, fail to reproduce, and die out over several seasons. Sometimes the poisoning is acute. Gill tissue is inflamed and damaged by a "spike" of extreme acidity, and the fish suffocate in excess mucus.

An entire "year-class" of newly hatched brook trout disappeared from the Paine Run watershed in Shenandoah National Park in 1992, almost certainly because of acidification, Bulger says. During the spring runoff of 1993 in the national forest a few miles to the west of Shenandoah, dozens of dead rainbow trout littered the streams, their gills loaded with aluminum. Meltwater from the heavy snows of the winter of 1996 was the most likely cause of death of the spring hatch of brook trout in the Saint Mary's. It was the second straight year that no new, young trout were seen there. A few reappeared the following spring.

Most of the scientific research on acid rain in the Blue Ridge and elsewhere has focused on sulfur dioxide emissions. A second group of pollutants, nitrogen and its chemical cousins nitrate, nitric acid, and the nitrogen oxides, has largely escaped scrutiny until recently. After all, nitrogen is often a fertilizer. It is taken up from the soil by most green plants, and it is essential for their growth. Folklore has it that farmers near the mountains don't have to spend as much on artificial applications of nitrogen because it falls out of the sky, dissolved in rainwater.

Human-made nitrogen emissions are mostly generated by electric power plants, industry, and auto exhaust. If it arrives during the growing season, much of the nitrogen is used by vegetation, so the soil, water, and aquatic life are less affected. But if it falls outside the growing season or in mature forests, where growth is slow, nitrogen can accumulate in the soil until it reaches toxic levels. For this reason soils and streams in older forests, such as those in wilderness areas, national forests, and national parks, are more likely to accumulate harmful levels of nitrogen than younger ones.

In a wilderness setting, even lesser amounts of nitrogen can set up a crazy quilt of artificially stimulated reactions altering, over time, the mix of species. Some trees grow faster with added nitrogen, but others don't. The fertilization effect may also stimulate some species to start spring growth too early or to continue growing too late in the fall, exposing them to damage from frost and cold. Nitrogen saturation slows down growth in some tree species. One study found that the total quantity of living plant material—"biomass"—and the general height of vegetation in a forest increased significantly with added nitrogen. Species diversity, however, *decreased* by a startling 60 percent.

So much nitrogen can be deposited on the soil that the plants—even if there are unusually high rates of growth—can't incorporate it all. When soils can't take in more nitrogen, another "saturated sponge" is the result. The excess nitrogen sheds into the streams.

Nitrogen also accumulates more rapidly in ailing forests that have been overwhelmed by disease or insects, such as gypsy moths in Shenandoah National Park. At high elevations in Great Smoky Mountains National Park where firs have been ravaged by the exotic adelgid insects, soils saturated with nitrogen have already been found. Streams there have the highest nitrate concentrations of any natural area in the United States.

Surplus nitrogen affects soil and water in many of the same ways that sulfur does. It diminishes their capacity to neutralize acidity, leaching aluminum and other elements, lowering the pH levels. And the effects of the two chemicals are additive.

"It's that old acidification scenario all over again, but it's based on nitrogen rather than sulfate," Forest Service research biologist Dennis Lemley says. "We're basically raising a red flag. We're seeing evidence of nitrogen saturation in the Smokies area."

When they are studying the effects of pollution, biologists can't monitor any ecosystem in its entirety—there's more of it than there are of them—so they select a few species, and a few places, to use as indicators of health or of trouble. This is like taking someone's temperature and blood pressure, rather than trying to examine every organ on every doctor visit.

If we hear news reports about these research sites—the Saint Mary's Wilderness, for example—or about "bio-indicator" species such as brook trout that are stand-ins for the ecosystems they are part of, it may seem as if they are the only organisms or the only places in jeopardy. But the brook trout's troubles, and those of the Saint Mary's River, are not unique. They are flickering vital signs, and they warn of declining health in watersheds all through the Blue Ridge.

If pollution diminishes enough in the future, the soils may recover at least to some degree. In Norway, experimenters built shelters over small watersheds, diverting acid rain and substituting unpolluted water. Thin soils, they found, regain their natural acid balance more quickly. Thicker soils, slower to reach acid saturation in the first place, also take longer to be rid of it. In many kinds of soil some sulfur pollution will remain, permanently.

Complete recovery for an acid-damaged ecosystem is unlikely. "In fact, rarely will distressed ecosystems return to their predisturbance condition after the cause of the disturbance has been removed, because the complex ecological interrelationships among predisturbance species are rarely the same," according to an EPA study.

No one can say, then, how completely the Saint Mary's and similar watersheds in the Blue Ridge might recover, but it is safe to assume that the less we poison, the more we can salvage. Perhaps, like me, you've never caught a brook trout and have rarely seen one. Maybe the disappearance of some bugs and water plants doesn't tug at the heartstrings. Understandable. But I like the cheerful way outdoor writer Ted Kerasote once sketched the picture:

"Most environmental problems . . . haven't affected, and probably won't affect, all of us in a direct way. But if they do—a favorite stream now fishless, a little bit of melanoma behind the ear—you'll feel it. And sooner or later some unwelcome industrial spinoff will touch almost all of us, which means that paying attention to one guy's acid rain may ensure that when your own little war heats up, you have some friends to call on."

The view west over the ridge from Slacks Overlook on the parkway on this late spring day is gray on gray. Haze, thick enough to induce claustrophobia, doesn't quite hide masses of standing dead oak and hemlock in the middle distance. Nearer at hand are dead dogwoods and some stunted remnants of the once dominant chestnut forest. Even the superintendent at nearby Shenandoah National Park has speculated that something's going on here that is not fully explained by exotic pests.

The conjunction of air pollution and ailing trees in the Blue Ridge and elsewhere has suggested cause and effect, and spurred scientific inquiry, for many years. But the role of air pollution in the widespread decline of trees—and whether there is an "unnatural" or unexplained decline at all—are strongly disputed, and the controversy is the source of frustration and suspicion.

News accounts since the mid-1980s have linked devastated forests and pollution. "Deadly Combination Felling Trees in East" read the headline on a page-one story in the *New York Times* in 1988. The view from these mountains, the story said, was "forcing many scientists to accept a conclusion that seemed only a possibility five years ago: air pollution, including acids in the air, is combining with natural stresses to cause heavy damage to forests throughout the East."

Most of the scientists who were quoted hedged their remarks, but their worry was evident. Several species including red spruce, fir, oaks, pines, and maple were suspected to be declining in health or in growth rate or both, because of pollution.

These alarms coincided with a ten-year, $550 million national acid rain study, sponsored by the federal government, whose results were published in 1990. Only a small fraction of the study was devoted to trees, but its findings were reassuring and seemed to contradict the scientists quoted in the earlier news coverage: "There is no evidence of an overall or pervasive decline of forests in the United States and Canada due to acidic deposition or any other stress factor."

It found that red spruce growing in New England was the only tree species clearly showing damage from acid rain. In the Blue Ridge, the evidence was said to be inconclusive. Widespread mortality among red spruce on Mount Mitchell was attributed to ice storms. But the report characterized itself as only "an initial exploration" and called for further investigations and careful monitoring.

The current research on air pollution's effects on spruce, oaks, and pines in the Blue Ridge merits a detailed summary. It serves as a case study of the sometimes awkward role that scientists and their research play in environmental controversies.

It seems a simple enough question to ask about the future. Is pollution killing trees, or not? We turn to scientists to track down the answer, but their work so far only raises other pressing questions: how highly do we value the mountain ecosystem, and how much risk will we tolerate? Those are issues that science cannot resolve.

"Effects" research is complicated by many factors—often only partly traceable—that influence trees during their long lives: genetics, climate, competition, diseases, insects, and prior disturbances such as fires and logging. Blue Ridge red spruce, though, seems especially well cast in the role of pollution victim. Most of it grows at 5,500 feet or more above sea level, on mountaintops and along high ridges that are cloud-bound for hundreds of hours each year, and the clouds are loaded with potential trouble.

The acidity of the mists poised on or blowing through the forest canopy has been measured near Mount Mitchell at an average pH of 3.7 and at times falls as low as 2.5—somewhere between vinegar and stomach acid. About 90 pounds of sulfur and nitrogen compounds are deposited annually on each acre of the thin soil by clouds, snow, dry particles, and rain. On Mount Mitchell itself, immersed in acidic clouds during more than a fourth of its summer hours, the figure reaches 120 pounds.

Plant physiologist Samuel McLaughlin of Oak Ridge National Laboratory says there is now far less uncertainty about acid rain and red spruce than a decade ago. "We've looked at the growth patterns of larger trees," he says. "That work has shown us that there is a slowdown in growth rate that was not explained well by climate or competition." That left air pollution or some other regional stress as a possible explainer, so his group followed up by tracing how acid rain affects the physiology of the trees.

"We found that acid rain leaches calcium from the spruce tissue and the soil, and can interfere with calcium uptake," McLaughlin says. Calcium is an essential nutrient for growth and for wood formation as a defense against disease.

Red spruce saplings were planted in soil taken from high elevations in the mountains, then exposed to carefully simulated acid rain. "It reduced growth and caused physiological changes in the trees that paralleled very closely what we have seen in the mountains," McLaughlin says. "We have looked at a number of factors that strengthen the evidence. You never prove things one hundred percent in natural systems, but you can do a whole series of studies that make it extremely probable that results are valid, and that they tie together."

But these results have been contradicted by still other research. One study, for example, found that after two years of exposure to both acidic

cloudwater and ozone, the growth of spruce seedlings was unaffected.
Shepard Zedaker, a Virginia Tech forest ecologist who has also been in the thick of the scientific debate about spruce trees and acid rain, says that some of his colleagues "have been crying wolf for a long time."

"We've been on the other side of the fence," he says. "Or at least, we can't sense an acid rain signal in all the noise in the data from other factors. The introduction of the balsam woolly adelgid into the forest, along with some tree-ring data from very old trees, lead us to believe that there's really nothing unusual going on in the Southeast."

At the highest elevations, only a few red spruce trees grow as individuals amid a forest of Fraser fir. When the firs disappeared, killed by the adelgid, the spruce were exposed to high winds and low temperatures. "When everything around you falls down, it's very hard to stand up to the rigors of that environment," Zedaker says. It may be, he adds, that southern red spruce trees are declining somewhat, but scientific proof of that and any connection with acid rain—if it materializes at all—is still in the future.

In the 1980s, forest ecologist Niki Nicholas helped gather growth-ring core samples from high-elevation spruce trees in the Smokies, which touched off the debate.

"Most researchers think acid rain has some effect on the forest eco-system," she says, "but they disagree on how strong the effect is." The most recent, and most complete, studies of the Smokies spruce trees, she adds, discern nothing abnormal in mortality rates, regeneration rates, crown condition, and growth rates.

Red, black, and scarlet oaks have also been put forward as pollution vic-tims. The source of abundant and nutritious acorns that are a mainstay for many kinds of wildlife, and the most common tree species group in the region, oaks are also in some of the worst trouble.

Nearly 70 percent of the oak stands in the Southern Appalachian re-gion's national forests are either already damaged or susceptible to a com-plex interaction of root disease, insect infestations, and drought called "oak decline." The Washington and Jefferson National Forests have been the hardest hit. About 20 percent of oak in the Pedlar District, in the Blue Ridge portion of the Washington National Forest, has been damaged. The trees die back slowly, from the top downward and the outside inward. Some recover, but others succumb in a few years after symptoms first occur.

Orie Loucks, an ecologist at Miami University of Ohio, and coauthor William Grant have written that sulfur dioxide, nitrogen, and, especially, ozone pollution are the "probable causal agents in the increasing rate of oak and hickory tree death, acting directly and indirectly," along with natural stress factors such as competition, drought, insects, and diseases. But as of 1998, their research had not been accepted for publication in a peer-reviewed scientific journal—an important step in gaining credibility among other scientists.

Other researchers have found little evidence of a tie-in between air pollution and oak decline. "There are some plausible explanations that are not as ominous or perhaps as politically charged," says Forest Service pathologist Steven Oak. The demise of the oak is instead, he says, the latest chapter in the recent natural history of the mountains.

The chestnut blight helped set up the situation when it removed a tree that competed with oak. When chestnuts died during the early decades of the century, oaks usually took their place. Oaks are currently common over huge areas where they may not be naturally suited to remain for long. They have enjoyed a short-term superabundance that is now fading.

Other disturbances have also played a role, Oak says. All but a small fraction of Blue Ridge forests have been logged, sometimes two, three, or four times, since the 1700s, and large areas were repeatedly cleared and burned for agriculture and hunting by both Native Americans and Europeans.

All these disturbances favored oaks, but they stopped sixty years or more ago: the chestnut blight has long since done its work; the burning, the clearing, and much of the logging have ceased. And the oaks that sprang up to replace chestnut trees are often dying now because of the "decline" factors. The weakened trees are not reproducing well, partly because other species are better competitors in low-disturbance conditions.

"I'm not going to say there's no effect from air pollution," Oak says, "but there's little evidence of it at the moment, and there are many other factors with documented effects." Other trees—black gum, red maple, sourwood, for example—will replace the oaks but not the acorn crop and the biodiversity the oaks have supported.

Southern pine forests, too, may be affected by acid rain, or maybe not. A recent exhaustive review of dozens of studies of air pollution and southern pines reached these conclusions:

- Acid rain, at current levels of acidity, probably does not affect the growth of southern pine seedlings and saplings. Even if acidity increased tenfold, it is unlikely to produce direct negative impacts.
- However, these conclusions hold only for young trees exposed to acidic rain for short durations. There has been no research on the long-term effects of acidic deposition on mature southern pine trees.

- Some soils are sensitive to increased acidity. They not only show decreased quantities of nutrients such as calcium but more metals, such as aluminum. Aluminum is not only toxic when dissolved but may interfere with trees' ability to take up nutrients.

"It will be extremely difficult to determine whether changes in soils are actually occurring and what factors are responsible if they are detected," another group of researchers concludes. But there may be "significant indirect effects" from acid on these forests, they warn, such as several cases in which exposure to acid rain has made trees more susceptible to attacks by insects and diseases.

As with red spruce, the research on pines sketches out some dire possibilities. Then it suggests that if stronger evidence of a link between acid rain and sick and dying forests exists, more time will pass before it is found.

Recent research tends to confirm the general picture of soils in some locales stripped of nutrients by acid rain, especially calcium and magnesium, which are essential for tree health. At a minimum, these findings tilt the seesaw of scientific evidence even further away from any justification for complacency.

In the soils of the Hubbard Brook Experimental Forest in New Hampshire, for example, acid rain has leached away large quantities of calcium and magnesium over the past quarter century, and vegetation growth "declined unexpectedly to a small rate" during the 1980s and 1990s. Aluminum dissolved by acid rain prevents soils from storing calcium, making less of it available to tree roots, another study found. And in a northern Pennsylvania forest of at least ten thousand acres, sugar maples have not reproduced in large stands since the 1950s. The mature maples are dying, with no satisfactory explanation. Ten tons of calcium per acre was added to the soil in one experiment, and the trees recovered significantly after five years. In other parts of the state, where acid rain is just as heavy but soils have more natural calcium and magnesium, maples are fine.

If you're uninclined to gamble with the health of the forests, such warnings are a call to arms. Paradoxically, they can also serve to rationalize a wait-and-see policy for those who insist on full scientific certainty. Calls for more research are easy to justify—more data is always a good thing—and from there it's just another half step, rhetorically, to argue against taking action.

The effects of another air pollutant, ground-level ozone, are quicker, plainer, and more widespread. Foliage bearing the visible signature of ozone pollution—mottles, stipples, flecks, reddening, dead surface cells, "burned" conifer needle tips, and other discoloration—is now a common part of the scenery all over the Blue Ridge.

Stroll one of the trails through the Big Meadows area of Shenandoah National Park in July or August and rangers can show you these effects on the needles of Virginia pine and Table Mountain pine and the leaves of yellow poplar, black cherry, sassafras, ash, and milkweed. They have been traced to ozone pollution.

Both laboratory studies and field observations have found evidence that ozone can cause the leaves of some kinds of trees to age more quickly, to fall off earlier, or to change color prematurely. Ozone may also dull some of the exquisite display of fall color that brings tens of thousands of visitors to the mountains each autumn, but those effects are difficult to sort out from other factors.

Field surveys in the Smokies have identified ninety native plants that show ozonelike leaf injuries, and about a third of those have been confirmed as ozone-sensitive. At higher elevations, as many as nine out of ten black cherry trees showed ozone injury in one survey in the national park, and up to three-fourths of the leaf area was affected.

Ozone marks the leaves and alters the chemistry of common milkweed plants, a staple of the diet of Monarch butterflies, and may affect their reproduction. There is "particular concern," according to the National Park Service, that ozone-induced chemical changes could make the butterflies more palatable prey for their natural enemies, birds.

Also known as photochemical smog, ozone is an unstable form of oxygen that combines relatively easily with the chemistry of other materials, including living organisms, in the process called oxidation. It's a source of confusion that in the stratosphere, 6 to 30 miles up, a layer of ozone is essential to protect the earth from incoming ultraviolet radiation. "Holes" in this ozone shield, caused by synthetic chemicals, are currently the subject of international concern.

Ozone also occurs naturally down here in the lower atmosphere, but human activities now add a far larger quantity of it to our sky, and not just in cities. Regional tides of air, thousands of square miles in extent and loaded with more than two or three times as much ozone as would occur naturally, flow over the Blue Ridge and surrounding rural areas during most of each summer.

Ozone is created when two classes of chemicals whose acronyms are VOCs and NOx (for volatile organic compounds and nitrogen oxides) react together in the sky under the influence of strong sunlight. Hot summer days, when the air stagnates, have the greatest potential for the creation of ozone. These are also the times when sulfur-based pollution is at its worst.

NOx in the Southeast comes mostly from coal- and oil-burning power plants, industrial boilers, motor vehicles, and other internal combustion engines. VOCs arise in roughly equal measure from vegetation and from human activity. As with sulfur dioxide, most of the ozone pollution in the northern Blue Ridge comes from the Midwest.

That ozone pollution alters the ecosystem is well known. It inhibits growth in a long list of agricultural crops; it makes two species of pines in the West grow more slowly or die. In the Blue Ridge, the case of certain kinds of white pines that were formerly common reminds us how much we have yet to learn about the ecosystem. An unknown number of genetic families within the species were especially sensitive to ozone injury and have probably died out as a result.

This may seem intuitively satisfying: an "only the strong survive" solution to the problem of air pollution. Unfortunately, though, this waste of biodiversity at the genetic level, which seems inconsequential in the short term, can prove to be significant.

As Auburn University's Arthur Chappelka explains: "When you lose genetic variations, you have robbed a resource from that particular species. For example, the white pines in that family were very fast-growing trees. And by narrowing the gene pool, you may set the species up to be vulnerable to a certain insect or disease."

Scientific proof that tells us whether ozone causes serious illness in trees has not arrived in a blockbuster announcement or two. It has instead accumulated slowly, as a hedged verdict. The fact that this pollutant leaves its mark on leaves and needles has been well documented. But those "burns" may have no more significance for tree health than a scatter of freckles does for the health of a child—a surface response to a largely benign environmental condition.

Instead, the growth rates of trees have become a central preoccupation in ozone effects research. For some, the threat of diminished wood production in the Southeast may come nearer the heart than faded fall color or troubled butterflies. It is also an indicator of forest health generally.

Phytopathologist John Skelly of Pennsylvania State University and his graduate students have carefully monitored ozone's effects on ash, cherry, tulip poplar, and white pine in the Blue Ridge for years, systematically gathering samples, sometimes teetering atop 100-foot ladders.

Skelly says there is as yet no proven connection between ozone's visible injury to tree leaves and needles and the growth of the trees. Too many other factors—moisture, light, heat, soil conditions, competition among trees, the intensity and duration of ozone exposures—confuse the data. If ozone is really affecting growth, he adds, confirmation is at least a decade away.

A summary of 159 research studies of ozone and vegetation, many of them conducted in the Blue Ridge, documents a slowdown in photosynthesis, sap production, root growth, and seedling growth in many species as a result of ozone exposure. But these results, not always repeatable and often tied to single species and particular circumstances, haven't added up to a final judgment either way. Taken as a whole, a recent report concluded, the research literature on ozone's effect on tree growth is inconclusive.

Others read the evidence differently, however, at least for pines. Another scholarly overview draws forth this conclusion from the pile of conflicting findings: ozone at current levels adversely affects the growth and health of southern pines and decreases their life span, photosynthesis, height, diameter growth, and mass.

So the "ozone question" turns out to be nearly the same as the "acid rain question": how much evidence of risk do we need, to justify taking action?

It is frequently suggested that the Blue Ridge ecosystem is caught up in a bad synergy—a series of mutually reinforcing natural and non-natural stressors—for which humans are the catalyst. The list includes acid rain, ozone, and recurrent waves of both exotic and native insects and diseases, whose attacks are said to be made more destructive by the already weakened condition of the trees.

"This may be an uncomfortable analogy for many, but the sickness the park has is comparable to a person with AIDS," former Shenandoah National Park superintendent Bill Wade says. "The rest of the thought is that because of the air pollution and acid deposition, a number of vegetation species, and probably the aquatic systems of the park—and who knows about animals?—are living with reduced ability to ward off other things that can impact them. Some of these are even natural events, such as drought, that they can withstand far less because of their reduced ability to do so."

Some research heightens the suspicion that destructive changes can combine as they undermine ecosystem health. A sampler:

- Hemlocks exposed to nitrogen pollution are more susceptible to damage from adelgid infestations.
- A study for the International Union of Forestry Research Organizations found "strong evidence" that chronic exposure to air pollutants such as ozone can decrease tree vitality and weaken resistance to insect damage. Several insect species, when presented a choice of foliage, almost always preferred the foliage exposed to pollution.
- A laboratory study of loblolly pine seedlings showed that elevated levels of ozone promoted larger fungal disease cankers.
- The dogwood anthracnose fungus responsible for nearly eliminating

that tree from the higher elevations of the Blue Ridge is more virulent
when seedlings are exposed to acid rain, which damages the outer
layer of the leaves.

But if single factors such as acid rain and ozone have been difficult for scientists to incriminate or exonerate, complex interactions are far harder to judge. As one group of scientists has written: "The concept of multiple stress was initially tempting, as it is vague enough to apply to any injury which cannot be explained in any other way. Its very versatility, however, makes it impossible to test and useless to those looking for the causes and mechanisms of forest damage."

Here the limitations of science are starkly apparent, caught between the magnificent, perhaps intractable complexity of the forest ecosystem on one side and our hunger for certainty on the other. Synergistic effects are not testable, these scientists are saying. Therefore, for scientific purposes, they cannot be acknowledged. Other researchers disagree, saying that multifactor forest ecosystem puzzles are a tough challenge but not an impossible one.

A bemused Arthur Chappelka says, "Sometimes I wish I worked in physics, where things are more straightforward. A lot of it is that we don't know. And it's hard for people to accept that. And they say well, gee, we should know. But if you think about this field, it's relatively young. Bacteriology has been going on for a couple of centuries. We've only been doing this about 20 years.

"It's a very complicated thing, and it's fun to work on, but it gets frustrating, because it's hard to explain to policy people," he says. Research results can't, for instance, precisely calibrate reductions in ozone pollution with increases in lumber production: "That's what they want to hear. And I can't say it."

Shepard Zedaker's outlook is similar, after long years of testing and sorting among the factors influencing forest health. "The epidemiology of forests is more difficult than the epidemiology of human populations as far as I'm concerned," he says, "because there are just so many confounding features."

The research agenda is shaped by the political agenda, Orie Loucks claims, in ways that often pose futile questions. "In the research about specific outcomes, like dead trees, studies of all sorts tend to narrow the scope of the questions. Part of what is happening is that people are being pushed and pushed to answer without reservation that there is a single factor. I call it the search for silver bullets. It is nearly impossible to think that you are getting the appropriate answer that way. Forests are an interaction among multiple species and multiple pollutants, not single ones."

It becomes clear, when the knotted intricacies of all this patient research are juxtaposed against questions about ailing forests, that the scientist's role is different from that of citizens and policy makers. Uncertainties about cause-and-effect relationships may linger for decades, as they did with smoking and lung cancer. Prudent decisions about soil and water, lungs and trees, ozone and acid rain cannot always wait for publication of the final research study.

At some point, these are matters of value rather than science, which John Skelly's attitude nicely illustrates. As a scientist, he is among the foremost skeptics about the effects of ozone and acid rain on the growth of trees. But as a citizen, he is indignant that ozone mars vegetation in supposedly pristine areas such as our national parks. The standard for regulating levels of pollution ought to be "no effects in the parks," he says, though imposing that standard would require large reductions in ozone emissions through-out much of the eastern half of the United States.

The quest for ever higher levels of scientific certainty on air pollution issues in the Blue Ridge and the rest of the Southeast can be a costly trade-off, a study by University of Virginia environmental policy analyst Vivian Thomson points out: "If we wait for greater scientific consensus on the many thorny issues, we thereby place a high value on scientific consensus and on minimizing economic impacts. . . . On the other hand, if we act on the available information . . . we are making a different value judgment: that precautionary stewardship for these natural areas takes high prece-dence in our political system."

This "precautionary principle," as it is called, acknowledges the need to act without scientific certainty at times when the stakes are too high to gamble.

Meanwhile, those who choose to wait for more substantial evidence of harm from air pollution will wait longer. State and federal budgets for ecosystem research have been cut repeatedly, and any scientist who studies air pollution effects can tick off several research programs that have winked out or whose budgets have shriveled. Long-term research can't be con-fidently planned, let alone carried out, in an era of now-you-see-it-now-you-don't funding.

Some Forest Service officials say that most of its surveillance is set up for evaluating trees as commercial timber, not as essential components of an ecosystem. The surveys are so intermittent and so broad-scale that more localized problems may well be overlooked.

One efficient and cheap solution is citizen volunteers. APHIS, the federal agency responsible for halting invading exotic plant pests, sometimes sets up 800-number hot lines. It encourages people to go outside, look for the bugs, and call in. Most of their finds turn out to be misidentified, but the system still reaches farther, faster, and with more efficient results in gathering "leads" than any other that the agency can devise on a limited budget.

In West Virginia, citizen volunteers have been trained to comb the national forests to check their condition. Most of their observations of sick trees turn out to have commonplace explanations, but not all. Forest pathologists there were astonished to find that a hitherto unknown disease of poplars had been turned up by a keenly observant amateur. The system also helps to defuse mutual suspicions between the Forest Service and environmentalists about "cover-ups" of dying forests.

The agency's official regional "forest health monitoring network," which doesn't involve volunteers, was established to collect the kind of long-term scientific data that leads to sound conclusions. Unfortunately, three of the five Blue Ridge states don't even participate. Virginia and Georgia are part of the program, but the Carolinas and Tennessee are not.

Also unfortunate: the network only samples 1 acre in every 43,000. One agency research forester says that "the sample intensity is pretty sparse, really—you might have one plot per county." Staff positions went unfilled and no data were collected in 1996 because of budget cuts.

There is another, nearer example of self-imposed blindness on the Blue Ridge. Climb the path to Devil's Courthouse, a granite fist bulging up out of the trees along the parkway south of Asheville. From here, your view on a sunny summer day should range out over Georgia, South Carolina, North Carolina, and Tennessee to a distance of 40 to 60 miles or farther. Now, though, the usual view on that kind of day is about 19 miles. All the rest smothers in haze. At times your vision penetrates only 8 miles, or less. Even Sam Knob, 2 miles away, looked pallid one recent June morning.

Ask a random dozen visitors and several might say it's the humidity, the heat, or vapor from the trees, which have always created the smoke in the Smokies and the blue of the Blue Ridge. That's partly true, and reassuring, because those are natural phenomena.

Others might figure that the view-killing haze has something to do with us, which is truer still. We know this, in scientific terms, only after decades of study. Human sensory organs have not evolved to perceive more than a fraction of the spectrum of life in mountain ecosystems. The minute patter of sulfate particles settling on trees, chemicals exchanging ions within the soil, the loss of fish populations in a watershed—these are inaudible, invisible, or too slowly wrought to sense directly. To know of them requires sustained attention and, often, plenty of technology—our artificial sense organs.

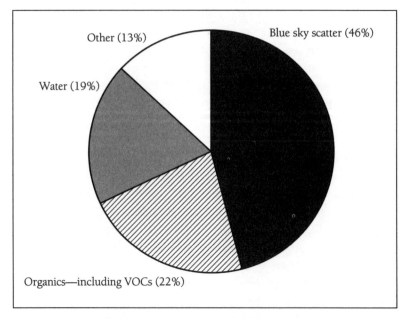

Figure 7. *Which kinds of natural air chemistry reduce visibility in the eastern United States?* These natural or "background" particles and gases cause only 10 to 30 percent of the atmospheric light extinction in the East. The figures are annual averages, and they do not include precipitation, blowing snow, and fog. (*Source*: Trijonis, "Natural Background Conditions," vol. 3, report 24, p. 76)

So an armada of instruments has been developed to measure the impact of pollution on visibility: nephelometers, diffusion scrubbers and denuders, transmissometers, single photon laser-induced fluorescers, and chemiluminescence detectors among them. They discern the patterns and trends, cross-check them, and pin them down.

For the impatient, the science here may seem to function as an ungainly machine designed for the splitting of hairs. But how else can we convince ourselves of what turns out to be the case: that milk-and-water shroud tinged with brown filling the near horizon is 70 to 90 percent human-made gases and particles, and they have thickened most rapidly just since midcentury.

Other than with somewhat stale collections of numbers, it's hard to remind ourselves what we've forfeited, because most of us rarely see it. The 284 overlooks that were the chief reason for building Skyline Drive and the Blue Ridge Parkway six decades ago were engineered as platforms for the magnificent views. They are now choked with haze, especially during summer months, when most visitors come.

In the words of the National Research Council, the pollution "reduces contrast, washes out colors, and renders distant landscape features indis-

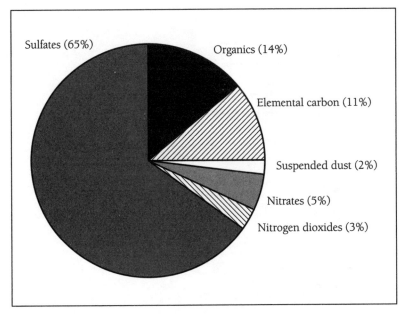

Figure 8. *Which kinds of human-made pollution reduce visibility in the eastern United States?* These pollutants cause 70 to 90 percent of "light extinction" in the East, including the Blue Ridge. The percentages are annual averages and do not include natural or "background" haze effects. (*Source*: National Research Council, *Protecting Visibility*, 213)

tinct or invisible." In fact, it can swallow even nearby mountain ranges whole (plate 6). At Shenandoah, where visitors once were able to see the Washington Monument, more than 70 miles distant, poor visibility is the number one complaint among park visitors. The view is affected by pollution at Shenandoah, the Smokies, and other park and wilderness areas in the Blue Ridge more than 90 percent of the time. On about a third of May-to-October days during the first half of the 1990s, visibility at Shenandoah was 10 miles or less (fig. 9). On average, two-thirds of the natural summer view in the Smokies is lost to pollution.

Visibility is diminished when, on even the clearest day, natural oxygen and nitrogen in the air deflect the light waves that bear a distant image toward our eyes. That's called "blue sky scatter." Natural sources of dust and gaseous emissions from vegetation also hinder light transmission. But these natural sources (fig. 7) account for only 10 to 30 percent of haze in the Blue Ridge. The rest is human-made (fig. 8).

Heat alone adds little to summer haze except for this: when we turn on our air conditioners in summer, demand for electricity goes up, so coal-burning utility plants generate more power and emit more sulfur dioxide. The haze intensifies.

The water vapor in the atmosphere—the humidity—doesn't create haze

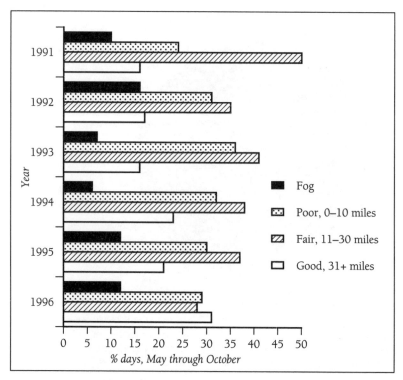

Figure 9. *Recent visibility at Shenandoah National Park.* The data are for May through October, the peak visitor season at the park. The view at the park is affected by pollution an estimated 90 percent of the time. (*Source*: Spitzer correspondence)

by itself, either. Instead, the water is absorbed by sulfate particles and they expand in size, so the view becomes even dimmer. (The "water" noted in fig. 7 refers to this combination of natural humidity with non-natural pollution.)

When the air was generally cleaner, the most visible air pollution was seen in plumes from industrial stacks, and it was assumed that haze came from somewhere nearby. Today, it routinely covers tens of thousands of square miles of the East—a mixed bag of pollution from a multitude of widespread sources.

On "dirty" summer days, 70 percent or more of the non-natural haze in the Blue Ridge is sulfur-based—the same pollutants that are the main components of acid rain, and from the same sources, principally coal-burning power plants.

Sulfur dioxide emissions declined in the United States during the 1980s and early 1990s—sharply in some places—and improvement in visibility was expected. But instead, research and monitoring showed that the sulfur concentrations in the summer air in reality trended upward at Shenandoah and in the Smokies during that period as a whole.

These data met with skepticism when they were first published. Some suggested that the monitoring instruments had malfunctioned or the numbers were somehow skewed. Special efforts were made to verify them, however, and they turned out to be accurate. Summer sulfur pollution was about 40 percent worse at Shenandoah in 1995 than it was in 1984, and 25 percent worse in the Smokies. After 1990, summer sulfur concentrations were stable at Shenandoah and may have declined, but only slightly, in the Smokies until 1995.

"In 1995, we had very poor visibility and very high sulfur levels," Great Smoky Mountains National Park air resource specialist James Renfro says. "Visibility got down to less than a mile. We were measuring visibility in feet on a couple of days in August. Cloudless days—no fog, no clouds, just the particles in the air." The following summer, 1996, was even worse, indeed the worst sulfur pollution on record at the park.

The good news may seem too simple: what technicians call "visibility impairment" lessens if clean air laws are strengthened and enforced conscientiously. Unlike some of the accumulating trouble that air pollution visits on soils, streams, and human health, the damage to visibility is reversible. Turn off the pollution and the scales fall from our eyes.

The pollutants that most effectively degrade visibility are small: about $1/200$th of the thickness of a human hair. If during some future July we stop sending them aloft, most of these particles will settle out of the sky or be washed out, in a few weeks or less. Nearly all the rest will be gone in a few months. Then, barring a thunderstorm that would bring a much sweeter kind of rain, we would see a far deeper, and bluer, sky.

cosystems intersect. Air pollution in the mountains finds its way to other organisms and other "sites of toxic action" than just trout gills and poplar leaves. It depends on species and circumstances. In your own case, some of the airborne particles you inhale—the street dust, the pollen, the first- or secondhand cigarette smoke, and the sulfur pollution—come to rest somewhere in your mouth, nose, and upper throat.

The finest particles, the ones that are less than a ten-thousandth of an inch in diameter, often penetrate more deeply into the respiratory tract. There, as it attempts to be rid of them, the body's clearance mechanisms are less effective. In the Blue Ridge during the summer, about 70 percent of the particle pollution in the atmosphere is of this "fine" variety. It is mostly those all-too-familiar sulfurous particles from coal-burning electric power plants—the same pollutants that generate acid rain and poor visibility.

Particles linger for minutes to hours in your nose and mouth until blown, sneezed, or coughed away or dissolved in your bloodstream or in mucus. But clearance of the fine particles—those that lodge in the ducts, sacs, and hairlike cilia in your lungs—takes weeks to years, if it occurs at all.

Your breathing patterns play a role, too. If you're hiking or cycling, breathing deeply or through your mouth, both fine and coarse particles travel farther down inside your lungs.

Our bodies have defenses against such intrusions. Concern about air pollution and health could be a kind of mild mental problem—an example of what some are pleased to dismiss as mere "chemophobia." But a forty-year-long trail of increasingly careful medical research links high levels of particle pollution, especially fine particles, with trouble for our species. Coughing, upper and lower respiratory infections, asthma symptoms, and changes in lung function become more common as fine-particle pollution intensifies.

If you already have respiratory and cardiovascular diseases, they are aggravated by fine-particle pollution. Hospital admissions, emergency room visits, school absences, and work loss days all increase with pollution concentrations. Epidemiological studies in the United States have consistently found a higher mortality rate in places where this kind of air pollution is worse. In other words, some of us—about fifteen thousand a year by the EPA's calculation—die sooner than we would if the air were cleaner.

A team at the Harvard School of Public Health tracked the medical histories of 8,111 adults for fourteen years in the eastern United States and accounted carefully for factors such as age, weight, income, smoking, and occupations that involve exposure to polluted air. They found that inhabitants of cities with the highest levels of air pollution had a 26 percent greater risk of premature death than those in cities with the lowest levels of air pollution.

Harvard and American Cancer Society research indicates that residents of the cities with the dirtiest air live, on average, two years less than those in our less polluted (but not at all pristine!) cities. And "average" means that the lives of many individuals are shortened by much more than two years.

The ailing, the elderly, and the young are said to be at greater risk. According to Dr. Ruth Etzel of the National Center for Environmental Health: "Children are more vulnerable to airborne pollution, because their airways are narrower than adults, they have much greater need for oxygen relative to their size, they breathe more rapidly and inhale more pollutant per pound of body weight, and they spend more time engaged in vigorous outdoor activities than adults."

What has all this to do with the mountains? Well, summer air in the Blue Ridge is rarely fresh. As a vacation refuge from befouled urban skies, it's an uncertain bet, better than the murkiest urban areas but not as clean as some others. "Park visitors don't realize that at times the air quality is worse in the park than where they came from," Great Smoky Mountains National Park air resource specialist Jim Renfro says.

Here's an optimistic reading of the statistics: under EPA standards current in 1998, air in the park may not have officially unhealthy concentrations of particle pollution.

Those EPA standards have been criticized by the American Lung Association as too lenient, though, especially for the safety of children, the elderly, and those with asthma or other breathing ailments. Using the association's recommended standards, fine-particle pollution reached or exceeded hazardous levels on more than half the summer days from 1993 to 1997 in both Shenandoah and the Smokies. Those are just averages. The worst days at the two national parks weren't merely unhealthy by Lung Association standards. They were "horrendously dirty," as a Park Service data analyst put it.

Ozone pollution poses still other problems for your lungs. If you've ever had to stop at odd times to catch your breath during a hike in the mountains, or while chasing your kids in a picnic area, it may not have been because you were out of shape. The high levels of ozone common in the Blue Ridge can be literally breathtaking.

Ozone concentrations in the atmosphere are measured in "parts per billion," or "ppb." The level of natural or "background" ozone varies, and it's the subject of some scientific disagreement, but higher estimates place it at something less than 30 to 50 parts per billion during summer. We humans add the rest, doubling or tripling the ozone in the air we breathe on a regular basis.

At levels around 80 ppb over several hours healthy adults suffer: reduced ability to breathe deeply, increased coughing and throat constriction, shortness of breath, pain upon deep inhalation, reduced ability to clear inhaled particles and microorganisms out of the lungs, changes in lung chemistry, increased respiratory tract inflammations in heavily exercising adults, and, perhaps, increased susceptibility to bacterial infection in the lower lung. The reactions of some asthmatics, who represent 5 to 10 percent of the U.S. population, can be far more severe.

Just as with fine-particle pollution, exercise worsens ozone symptoms. When you inhale it, ozone reacts with the fluids, mucus, and cell linings of your respiratory system, and exercise makes you inhale more deeply and frequently. For most of us, fortunately, the symptoms will subside in one to several hours.

Repeated exposures, however, may bring a different result. "When you get sunburned," EPA medical officer William MacDonald says, "you get pain and redness, and if it's bad enough the dead cells slough off and you develop a new layer of cells that cover the skin. If you do this over and over again during a number of years, your skin gets thicker and loses elasticity."

He suspects something similar—a kind of "sunburn" on the lung tissues—may occur with repeated exposure to ozone. A good brisk hike over about six hours in the Blue Ridge when the ozone is at 80 ppb damages some of the cells lining the lung. "It also causes redness and pain," MacDonald says. "If the damage is severe enough, some of those cells slough off in a few days. They're replaced by new cells, and this is the part we don't know: if you go through these cycles of damage and repair, I'm concerned that, as with the skin, you may get some sort of chronic changes in the lung."

The EPA says that the ozone "safety limit" is an average of 80 ppb over eight hours. Keep "80 ppb" in mind, and the following numbers give a fair indication of the extent of the health problem in the Blue Ridge:

- By some measures, the Cove Mountain monitoring station in Great Smoky Mountains National Park has the worst ozone pollution of any rural site in the United States, and it may be getting worse. It exceeded the eight-hour, 80 ppb EPA health limit sixty-two times during 1993–96.
- At the other end of the Blue Ridge, Shenandoah National Park's ozone levels would have exceeded the same standard in four of the same six years. Monitoring stations recorded nineteen days with eight consecutive hours of ozone levels higher than 80 ppb during the first five years of the 1990s. One-hour levels up to 120 ppb have been recorded.

- Roughly halfway between the two national parks is White Top Mountain, part of the Mount Rogers National Recreation Area in Virginia, adjacent to two federal wilderness areas and Grayson Highlands State Park. Eight-hour averages were not available, but one-hour ozone levels have reached 163 ppb on this remote peak, and the maximum readings were 101 ppb and 108 ppb in 1994 and 1995. In each of those years, ozone levels exceeded 80 ppb for a total of more than a hundred hours.

Restoring clean air in the Blue Ridge won't be free, but it has gotten cheaper as time passes. Utility industry estimates of the cost of removing a ton of sulfur dioxide from emissions tumbled 95 percent between 1990 (when the utilities were using the estimates to lobby against air pollution standards) and 1996 (when the standards had been in place for six years). Costs could rise again somewhat in the future.

So is it worthwhile to get rid of air pollution, or not? For some, that's an economic question. The EPA estimates that each American paid an average of about four dollars a year during the first half of the 1990s to reduce sulfur dioxide emissions. Alas, the rule of thumb is that as more pollution is scrubbed out of tailpipes and smokestacks, each increment is more expensive than the last. Would you pay twenty dollars a year instead?

Economists disagree as to whether the cost of the cleanup is more or less than the market value of the technological advances it generates. World market leadership in new antipollution technology is, for instance, a great economic boon. Another calculation is even more vexed: how much is a brook trout population, a swarm of the mayflies they feed on, a healthy watershed, a clear sky, or a set of unimpaired lungs worth to you?

The EPA has carefully estimated the costs, for victims of sulfur pollution, of lost work time, hospitalization, medicine, and inability to perform household tasks. Also considered in its calculations were "quality of life" issues: the amount people said they'd be willing to pay to avoid reduced enjoyment of life, pain, suffering, anxiety about the future, and concern and inconvenience to family members and others.

Other research was incorporated in the final dollar figure: people were asked what they would be willing to pay to avoid "persistent symptoms of cough and phlegm, limits on physical activities, and ongoing medical care" from chronic bronchitis. Unsurprisingly, those who are on more intimate terms with this question because they or their relatives have emphysema, asthma, or chronic bronchitis tend to be willing to pay more.

The EPA study decided to go with a quarter of a million dollars as the middling figure that people say they would be willing to pay, for all reasons, to avoid an average case of chronic bronchitis; $14,000 to avoid a hospital visit for cardiac or respiratory problems; $2 million to $7 million to avoid a "premature death" (see endnotes).

On the basis of those figures, the EPA concludes that the yearly benefits to health of clean air legislation approved in 1990 will outweigh its costs by anywhere from 5–1 to 16–1 by the year 2010.

The assumptions that underlie such price tags always provoke debate. One set of critics of the Clean Air Act—economist Ben Bolch and chemistry professor Harold Lyons of Rhodes College, Memphis—demands to know why "those estimates [of benefits] never seem to include the death that might be caused by loss of income from such draconian controls. . . . Could it be that the numbers of lives saved from emission control might be less than the number lost because of reduced incomes for the poor?" Their intriguing equation—more pollution could mean longer-lived poor people—awaits further research.

Air quality in the Blue Ridge will depend during the new century on how well we keep certain promises we have made to ourselves in various formulations of the Clean Air Act. These IOUs have come in three waves—for simplicity, let's call them "Phases 1, 2, and 3"—roughly in the 1970s, in 1990, and in 1997. On each occasion, new controls over various kinds of pollution that affect human health, visibility, and the natural environment were to be implemented slowly, over 10 to 20 years. This schedule acknowledged the expense and difficulty inherent in reengineering, for example, automobile engines and electric power plants. It also offers plenty of time for postponements, indecision, and retrenchment.

Despite each set of new plans, and the aura of relief and self-congratulation that went with them, the debate over air quality did not end when they were approved. Negotiation, interpretation, litigation, and political pressure determine how and whether air quality laws are effective, and whether they are really enforced or neglected. The results so far have been promising, uncertain, equivocal.

Phase 1

The first air quality legislation, back in the 1970s, set out important goals that proved difficult to realize. Visibility improvement was one such goal and can serve as a rough index of how the cleanup efforts have fared. Visibility has been tracked with special care in the two national parks and six of the wilderness areas in the Blue Ridge, because Congress declared at the time that these were part of a new set of "Class I" areas (fig. 10). It established as a national goal "the prevention of any future, and the

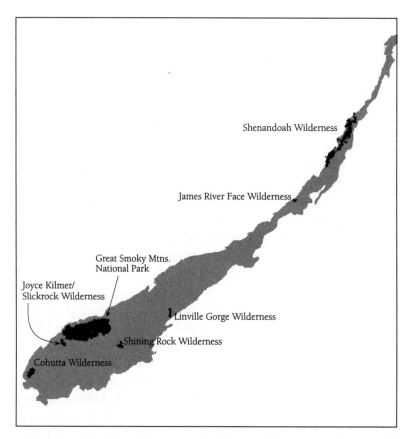

Figure 10. *"Class I" airsheds in the Blue Ridge.* Federal clean air laws are supposed to afford maximum protection for "Class I" areas, but they have had only limited success in halting the development of new sources of pollution nearby. Air pollution from more distant sources blankets the whole region, including "Class I" areas. (*Sources:* Keys et al., *Ecological Units of the Eastern United States,* CD-ROM; Hermann, *Southern Appalachian Assessment GIS Data Base,* CD-ROM)

remedying of any existing, impairment of visibility at these sites." But we view them now as we did then: dimly, through cataracts of pollution.

The National Research Council, which advises the federal government on scientific and technical matters, was asked to investigate the lack of progress after fifteen years. It declared that no additional scientific research or technology was needed in order to take action. The problem: "lack of commitment to an adequate government effort."

Phase 2

In 1990, before much of Phase 1 was fully implemented, new amendments to the Clean Air Act were passed by Congress, to be applied in stages until the year 2010. By now, some of these air quality regulations—"Phase 2"—may have been superseded. On the other hand, similar rules could still be in place, so it's important to know what they would achieve.

The Phase 2 rules would rescue only a fraction of the visibility we've lost, for instance. In the Smokies and at Shenandoah National Park, more than half of the natural visual range would still be obscured by pollution in the summer of 2010, when the rules were to be fully implemented.

Ozone pollution was even less likely to diminish. Some research has found that future population growth, increasing demand for electricity, and an ever-denser swarm of auto traffic will increase both VOCs and NOx—the compounds that mix to create ozone—despite Phase 2 rules.

One EPA computer model generated better news for the Blue Ridge, however. It predicted that high-ozone episodes would be markedly fewer by about the year 2005, especially in the area bracketed by the James River Face Wilderness and Shenandoah National Park. But these data also indicate that the highest ozone levels along nearly the whole span of the Blue Ridge would still exceed 120 parts per billion at times.

If so, ozone would remain far above "natural" levels in 2005, one government analysis points out. Our national commitment to leave the parks "unimpaired" for future generations, and to protect the wildernesses so that "the influence of people is substantially unnoticeable," would still be a distant goal. And Blue Ridge ozone pollution broke records in 1998.

By the mid-1990s, there were reports of significant decreases in some kinds of sulfur pollution descending on certain areas of the Blue Ridge as Phase 2 took effect. Other indicators and monitoring sites seemed to show that the level of pollution was holding steady or even getting worse. In 1996, monitors at Great Smoky Mountains National Park measured the worst summer for fine-particle sulfate pollution since records were first collected, in 1984. Compared with that year, 1996 summer sulfate levels at both of the national parks had risen more than 20 percent.

The EPA's predictions of the coming century's sulfur-based pollution involve a fairly wide range of uncertainty. Things will improve as the dirtier, older electric power plants that pump most of this kind of trouble into the atmosphere are slowly phased out or refitted with cleanup technology. The question is, how long will it take?

Pollution control laws are more lenient for the older, dirtier plants, and there are few other economic reasons to retire them from use, since they have already been paid for. "Together these factors have created an incentive for plant operators to keep existing plants in service as long as possible," a government report notes. "Life extension has become a priority for the industry, which is seeking low-cost approaches to keeping up with increasing demand. . . . Although it is expensive, it is economical compared with new plant construction." Relying in part on that record, EPA projections assume that under Phase 2, sulfur pollution would not decline after 2010—and nitrogen pollution would increase.

Scientists concluded that Phase 2 wasn't strict enough to stop the trend of increasing acidification of soil and water in the Blue Ridge.

The EPA predicted that south of Blowing Rock, North Carolina:

- A handful of streams would become chronically acidic by the year 2040. In those cases, the "sponge" would be saturated, so all incoming acid would go directly into the streams.
- By that year, from 11 to 16 percent of the streams would suffer acute acidification during snowmelts and heavy rains, which often come during the spring hatch when fish are most at risk.

A different set of projections by a team of University of Virginia researchers concentrated on the more acid-sensitive half of the Blue Ridge from White Top Mountain in southern Virginia to the northern tip of Shenandoah National Park. They concluded that by the year 2050 under Phase 2 rules, brook trout would sicken in and disappear from a third of the streams they now inhabit. Remnant populations might or might not survive in the area. University of Virginia fish physiologist Arthur Bulger's forecast: Phase 2 "means that things are still going to be getting worse, but they're going to be getting worse more slowly."

Phase 3

By the mid-1990s, Phase 2 began to weaken. Loopholes, lobbying, and political pressure accumulated at the federal and state agencies that interpret and enforce air quality laws. There was great uncertainty in those agencies about whether Americans really support antipollution measures enough to defend their enforcement.

Top federal and state officials declined to back up the administrators at Shenandoah and Great Smoky Mountains National Parks when they tried to forestall new industrial plants from adding to pollution in the park's airsheds. Congress was considering other changes to cut back on air quality requirements. Citing alleged public opposition to the expense and inconvenience, state governors battled the EPA as it tried to implement limits on auto-exhaust pollution.

At the same time, however, the EPA was being pushed in the opposite direction, toward stiffer air quality regulation. The agency was successfully sued by the American Lung Association for not enforcing its own air quality laws. Scientific evidence, linking fine-particle pollution and ozone pollution to ill health, mounted.

So in 1997, a set of new regulations we'll call Phase 3 was announced by the EPA. One part of this package imposes tougher health standards for both ozone and fine-particle pollution (which includes sulfur-based pollution).

Researchers did not know whether these new provisions might remove enough pollution to halt the ongoing acidification of soils and streams in the Blue Ridge. They would, beyond question, help a lot if they are enforced.

The timetable for enforcing the new rules is a stretch, however: areas that don't meet the new ozone standard may have until the year 2012 to do so. Areas that don't meet the new fine-particle pollution standards may have until 2017.

New visibility standards for "Class I" areas (fig. 10) were also due from the EPA as the end of the century neared. The National Park Service found a draft version rather too forgiving. At the proposed rate of improvement, it would take two or three hundred years for natural visibility to be restored to the Blue Ridge.

But Phase 3, which has generated ample political opposition, may not be carried out at all. During the year or so after the new ozone and particle pollution standards were announced, opponents vowed that they would be repealed or softened, and Congress considered a bill that would impose a four-year moratorium on implementing them. Governors in some of the Blue Ridge states were talking of a "revolt" and petitioned the federal government to reconsider or delay some of the new requirements.

Deregulation of electric power utilities is under way, too, producing increased competition and strong incentives to continue using dirty, coal-burning power plants to generate cheap electricity. Research on the effects of deregulation by a consortium of northeastern states has made them "fearful," their report says. They worry that sulfur-based pollution will increase, not decrease, in the future, despite Phase 3.

Neither supporters nor opponents of this latest set of clean air rules can regard it, then, as a "done deal." It is, instead, an ongoing tug-of-war. By the year 2020 or so we will know—in the Blue Ridge and elsewhere—who pulled hardest.

M ount Oglethorpe, the original terminus of the Appalachian Trail, is the southernmost peak of the Blue Ridge. It tops out at 3,290 feet in a radar installation, a grove of scrub oaks strewn with beer bottles and a cracked, graffitoed monument to colonial Georgia's founding father, James Oglethorpe.

The view south is over a field of stumps. One of them, at least when I was there, wore an inverted pair of men's boxer shorts ablaze with hearts and the legend HOT STUFF.

Other signs of the vigor of the human influx abound. Get-away-from-it-all vacation homes ascend the ridges, bringing a generous portion of "it" with them. Bare, eroding soil heralds dozens of new building sites on steep slopes nearby. "The North Georgia mountains have been waiting for you since the beginning of time," one sales brochure explains.

New subdivisions are also springing up along the boundaries of Shenandoah National Park, especially in the northern district. Chain saws, lawn mowers, dogs, and loud music can be heard within the slender park, even in designated wilderness areas, where solitude is of the essence. Park resource manager Steve Bair says housing construction has "really been wide open here lately." New homes at the boundaries number in the dozens now and will soon be in the hundreds, he predicts.

"In 50 years there will be some very steep mountain land that won't be developed, but I have the feeling that any fairly flat areas that can be built upon, will be. It's going to happen, and I think we'll be pretty much an island in a sea of development."

Greenbelts, dozens of miles deep, of private woodlands and farms dotted with occasional hamlets used to extend east and west all along the Blue Ridge. They insulated the national parks and forests from the effects of suburbanization. Now, increasingly rare and more prized, these buffers are being roaded, subdivided, sold off, and made rarer still. They are being replaced by a spreading sequence of homes and commercial strips that replicates any other American urbzone.

Like a field biologist's transect string stretched across an observation plot, the celebrated view on and near the 469-mile Blue Ridge Parkway is an index of the changes in the mountains. Heading south near the start of Virginia's piece of the parkway, you can sense the wisdom in how the road's gentle curves were laid, seven decades ago. They guide sight lines out over expanses of forest and far-off valleys. Then abruptly, the hard geometry of a four-story ski condo juts from a high ridgetop, like a machine on the sky.

Farther on, in Virginia's Washington National Forest, is Crabtree Falls, a series of cascades that hurtle and crash down a thousand feet of rocks. From the promontory where a trail begins, and as you descend along the

SOLUTIONS 8

A House in the Mountains

Blue Ridge Parkway landscape architect Will Orr produced this "top ten list" for building sensitively in a scenic mountain landscape:

- If you are building on a ridge, place your structure on either side of the ridge, not on top. Placing a structure on top of a ridge may provide increased views for you, but it also increases the view of your structure. In addition, the extreme weather of the Blue Ridge will work to tear your structure apart if it is not buffered from the wind. This will increase your maintenance. If possible, place your structure on a south-facing slope.

- If you are building on open land, place your structure at the edge of the pasture or meadow. It is human tendency to want to fill open spaces. These openings should be preserved. Instead, place your structure at the woods edge. This will allow your structure to become a part of the scene. Other benefits include reduced amount of tree clearing, adequate buffers for your structure from wind, and shade in summer to reduce cooling costs.

- If you are building on a forested slope, retain as much forest cover as possible to screen your house or development. In order to create a view out from your house, selectively remove individual trees. Remove lower branches from remaining trees. This will create a canopy view that will provide a more interesting view and will be easier to maintain in the future.

- Construct driveways that gradually climb the slope. Avoid driveways that go directly up a slope. Many new houses today have driveways that run straight up the mountain. These driveways look unnatural, are very difficult to maintain, and are hazardous during winter. Instead, construct a driveway that goes up the slope gradually at no more than a 15 percent slope. This slope can be defined by a walk that makes it

falls, the view is now altered. A door into the wilderness has been left ajar, and the suburbs are tumbling through. New white and blue and red houses jump out from the deep green swells of the opposite ridge.

The Flat Rock loop trail off the parkway in North Carolina leads through the woods and out onto a kind of rock plaza, affording a long view up the flanks of Grandfather Mountain, from which prominent sets of living-room windows return your gaze. In the foreground, a carpenter's hammer barks into the afternoon from the fast-growing Linville community.

Well south of the parkway, on South Carolina's sliver of cool, forested somewhat hard to breathe but not impossible to walk without stopping. This drive will take more land to construct and cost more but will be worth the initial investment.

- Use building forms that mimic old architecture. Appalachian architecture generally incorporates the following elements:
  One-story houses with basements cut into a hill.
  Wood siding—the texture created with wood can be copied through several modern materials that are easier to maintain than wood.
- Avoid structures that appear as a large, single block. Break your structure into smaller pieces. These pieces will help your structure to blend into the site. This type of architecture will also create a home that is more intimate.
- Construct porches. Porches allow the structure to blend into the natural environment. In addition, porches provide wonderful outdoor spaces that can be used nearly year round with the climate found on the Blue Ridge.
- Use warm gray colors for stains, paints, and roof materials. The color of warm gray is defined as gray with brown mixed in it. This color can be found on weathered rock faces and tree trunks. Gray will blend into a scene much easier than browns or greens.
- Use architectural-grade asphalt shingles. These shingles are thicker and will thus last longer. These shingles also come in "multi-tab," which helps to provide a texture similar to wood shakes but without the maintenance concerns.
- Avoid large panes of glass for windows. Glass is very reflective and can make a house on a mountain slope very visible. If you want to provide a large opening for looking out, consider a bank of smaller windows. This is also in keeping with regional architecture.

highlands at the edge of the east-facing Blue Ridge escarpment, is a 3,100-foot granite hump called Glassy Mountain. Some had hoped that Glassy would become the centerpiece of a park. Nine hundred building parcels and a golf course now cover 3,000 acres at the top. Two restaurants, a chapel, a pavilion, and a heliport are part of the plan.

"We're having huge rates of change," says the parkway's landscape architect, Will Orr. "The prime retirement spots used to be Texas, Florida and California. Virginia and North Carolina are the number one retirement places in the country now. The mountains of Appalachia have been rediscovered. Lots of people are moving in. Lots of second homes. And of course the construction trend right now is not small two-bedroom cottages, but the 3,000-square-foot three-story big old humongous building."

Though it is the most-visited unit in the national park system, the parkway was designed to handle traffic and is far from reaching its capacity except during the autumn leaf season or near urban areas such as Roanoke and Asheville. Because it spans four national forests, many of the views, so intuitively engineered, will be preserved.

But as you trace a finger along the route, areas where the surge of future development may replace the natural landscape are apparent. This "linear park" is skinny, averaging only 500 feet of protected area on either side of its center line, and much of it is flanked by private land. There, development will continue to gallop, especially in southern Virginia and northern North Carolina. "Once you get to Roanoke, we're likely to have a build-out through that whole valley," Orr says. Its gently rolling topography, well suited to farming, also invites a rising tide of construction and commerce.

Building along the steeper slopes farther south will be even more likely to alter the scenic character of the road. "On a mountainside, you're going to see that house, and it's going to really jump up and fight the scenery. There's no good solution as to how to hide a house on a site like that," Orr says. The already robust catalog of examples grows. At Orchard Gap, near milepost 195, plainly visible, evenly spaced chalet A-frame vacation homes turn a ridgeline into a sawtooth of Swiss-style rooflines. Moses H. Cone Park, along the parkway near Blowing Rock, looks out over Bass Lake. The blank facade of an apartment building, denuded of the saving grace of trees or landscaping, looms from the opposite ridge.

Growth is inevitable. How it is managed will determine the character of the parkway, and much of the rest of the Blue Ridge, in the near future. Builders, architects, planners, and citizen groups in Roanoke have cooperated to soften the visual impact of some new housing construction there, for example. Their work is a model for other mountain communities as they try to cope with a burst of new development.

But Boone, North Carolina, is a different kind of model, a development hub whose already speedy expansion will be fueled by planned new state highway projects. Most flat land in the area is already a carpet of stores and homes. They are now climbing the slopes. "Great views from those windows," Orr says, "but the house sticks out like a sore thumb."

Haphazard growth threatens to "junk up" the mountains, Orr says. "I'm afraid Boone is a good example of what's going to happen when you have a big thrust of people without planning the place." The 170 green miles of the parkway from Roanoke to Blowing Rock, most of them pri-

SOLUTIONS 9

More than Just a Road

Biologists have recommended that these steps be taken to protect rare and endangered animals along the Blue Ridge Parkway:

- Hire naturalists, research coordinators, and people who can interpret the natural assets of the parkway for the general public.
- Raise salaries and improve working conditions for such employees, so the exceptional and experienced ones will want to stay longer. Rapid turnover is a problem.
- Too much development to accommodate the automobile or provide the amusements and amenities found in nearby towns will tend to cut the preservation of wildlife habitat along the parkway.
- The parkway should consider whether there is an upper limit on cars, people, abandoned cats and dogs, and human disturbance that needs to be considered in light of the vulnerability of species and habitats. Can popularity ultimately destroy the aesthetic, biological, and economic benefits of a park?
- More monitoring and research on the parkway's natural resources is needed.

Source: Knowles, Steele, and Weigl, "Rare and Endangered Vertebrates," 152–54.

vately owned, could be lined with houses and strip malls in another thirty years, he adds. Farther south, areas such as Linville, Spruce Pine, Little Switzerland, Balsam Gap, and Waynesville are also ripe for a suburban metamorphosis.

Documenting such growth can be elusive. Because detailed community planning efforts are rare, there is no compiled record of the large tracts of forest and mountain land that are continually subdivided into smaller parcels or of the roads punched in to provide access.

In an effort to fathom how the landscape may be subdivided in the future, planners in Virginia surveyed Greene, Albemarle, Nelson, Fluvanna, and Louisa Counties and the city of Charlottesville, all in or near the Blue Ridge. Steep slopes, flood plains, areas where septic tanks won't function, and other large sections of land were eliminated as possibilities for future development.

Their work shows, senior planner Michael Collins says, that local ordinances currently allow "nearly a complete conversion from rural to suburban land uses." As farms and forests are atomized, the study area's current

Table 2. The Sixteen Fastest-Growing Blue Ridge Counties

	Percentage Change, 1990–95	Average Percentage per Year
Dawson (Ga.)	29.8	5.96
Greene (Va.)	22.9	4.58
Union (Ga.)	20.3	4.06
Gilmer (Ga.)	19.1	3.82
Johnson (Tenn.)	18.7	3.74
White (Ga.)	17.9	3.58
Pickens (Ga.)	17.0	3.40
Sevier (Tenn.)	16.7	3.34
Lumpkin (Ga.)	15.4	3.08
Murray (Ga.)	14.9	2.98
Bedford (Va.)	14.6	2.92
Blount (Tenn.)	12.8	2.56
Towns (Ga.)	11.9	2.38
Warren (Va.)	11.9	2.38
Macon (N.C.)	11.4	2.28
Henderson (N.C.)	10.7	2.14

Source: U.S. Bureau of the Census.

population of 173,000 could eventually more than quintuple, to 993,000. In 1990, there were 72,000 houses in the area. The eventual "build-out" could be 432,000.

Census Bureau data on population growth in the mountains from 1990 to 1995 strongly signal the current trend.* During that period an estimated 200,000 additional people established permanent residence in the seventy counties along the Blue Ridge.

About a third of the counties grew more than 9 percent during just those five years, a rate that University of Georgia population expert Douglas Bachtel calls "phenomenal." Nine counties grew more than 15 percent, and three more than 20 percent (table 2). By contrast, growth throughout the United States during the same period averaged 5.6 percent.

*In 1996, a campaign was under way in Congress to abolish the Bureau of the Census.

"They call them 'half-backs' up there," Bachtel says. "What they often
have is people from up north who moved to Florida. Then they get sick of Florida, and they move to Northeast Georgia, and so they're considered to be 'halfway back' home."

A new birth doesn't affect a community and its natural environment right away. But the impact of an adult's arrival on housing, roads, and the need for services is immediate. In Georgia's Union, Rabun, and Towns Counties, more than 90 percent of the growth was from new people moving in, rather than from local births. The mountain populations, Bachtel says, "are just growing like crazy. There ain't no doubt about it."

But the highly visible new houses moored like barges on nearly every ridge facing Whiteside Mountain near Highlands, North Carolina, or perched atop eroded scars along Highway 74 in Tennessee are often invisible in the census numbers. Incomplete, the census masks the real face of human presence in the Blue Ridge, because it does not include most seasonal residents. The development boom in second homes, vacation homes, and part-year homes for retirees doesn't register.

These changing scenes reflect our aspirations, as well as our numbers. We seek the serenity and the beauty of natural areas so avidly that, soon, it's gone, "loved to death."

Consider the odyssey of Alfred Bartlett. "When I moved to South Florida from Pittsburgh in '59," he recalls, "we thoroughly enjoyed it there. My wife and I both grew up in the city, and when we moved to Fort Lauderdale we could go down and actually see the beach."

Around 1980, the Bartletts built a vacation home on Beech Mountain, a half hour from Boone. The summers were cool at 5,500 feet, there was a campground on top of the mountain, and, perhaps best of all, it was different from Florida. By then, Florida had changed. The natural beauty of the coast had been replaced by a wall of buildings.

"It was all gone, mainly because of the population," Bartlett says. "And people would think, gee, why did they do that? Why did they allow the buildings and construction and population to go into those areas to that degree?"

The Blue Ridge offered what Florida had lost. It wasn't as crowded, not so pushy. The pace was a lot slower and the people were friendlier. The Bartletts had to drive down to the bottom of Beech Mountain for groceries then, to a tiny store that doubled as a meat market.

Twenty years later, they have their choice of three supermarkets nearby. They have added on to their house four times. The campground on top of the mountain is gone, replaced by condos.

"Everything has probably doubled or tripled in size," Bartlett says. "The

SOLUTIONS 10

The Future of Sprawl

[In California, the Bank of America and a coalition of other interest groups produced a report on the effects of, and the future of, suburban sprawl—a glimpse of the future, perhaps, for the Blue Ridge. Edited excerpts:]

This acceleration of sprawl has surfaced enormous social, environmental, and economic costs, which until now have been hidden, ignored, or quietly borne by society.

We can no longer afford the luxury of sprawl. . . . We cannot shape the future successfully unless we move beyond sprawl. This is not a call for limiting growth but a call to be smarter, to invent ways we can create compact and efficient growth patterns that are responsive to the needs of people at all income levels.

Continued sprawl may seem inexpensive for a new homebuyer or a growing business on the suburban fringe, but the ultimate cost—to those home-owners, to the government, and to society at large—is potentially crippling.

As fiscal and cost-benefit analysis techniques have become more refined, the true cost of sprawl has become much more apparent: the cost of building and maintaining highways and other major infrastructure improvements to serve distant suburbs; the cost of solving environmental problems (wetlands, endangered species, air pollution, water pollution) caused by development of virgin land on the metropolitan fringe.

A do-nothing approach, in effect, constitutes a policy decision in favor of the status quo. While the state and the regions have created a leadership void in this area, many local governments have stepped in with their own pol-icies, which often have served to promote sprawl rather than prevent it.

roads, the number of businesses, and the traffic have increased tremen-dously. More and more homes are built every year. We have a lot of folks up here that have discovered this area like we did, and so there are a lot more people."

The Bartletts used to take their children to Sliding Rock in the Smokies, where they could zip down into the pool at the bottom of a granite chute. "You go back to those places today . . . just lines and lines and tons and tons of people. We talk about it. We think about the way things used to be. I'm sure the folks who were born and raised here think about it a lot more than we do."

The natural surroundings on the mountain are still beautiful, but as in Florida, so in the Blue Ridge. "That's exactly what's going to happen here. People tend to forget what they came for. The only thing I can see is to enact some kind of law that would prevent that sort of thing."

Regardless of the methods used, much of the leadership for providing greater certainty for conservation and development must come from the state, regional agencies, and local governments working together. But private businesses also have a critical role. New real estate developments can be brought to market more quickly and cheaply within areas where effective consensus plans for conservation and development have been created.

This report calls for collaborative efforts by all sectors of society to work on ways of lowering sprawl's adverse social, economic, and environmental impacts while encouraging strong economic growth.

Such efforts should be guided by four goals:

- Provide more certainty in determining where new development should and should not occur.
- Make more efficient use of land that has already been developed including a strong focus on job creation and housing in established urban areas.
- Establish a legal and procedural framework that will create the desired certainty and send the right signals to investors.
- Build a broad-based constituency to combat sprawl that includes businesses, environmentalists, community organizations, farmers, government leaders, and others.

Source: Bankamerica Corporation, "Beyond Sprawl."

The Bartletts have thought of moving. "I'm looking," he says, "for a place that is what this used to be. If that place exists, I don't know where it is."

Wherever they may be, such quiet places are transformed by roads. In the flatlands, what was once our high-speed, ultraconvenient interstate highway system is now, often, a knotwork of clogged arteries. Throughout the Blue Ridge more and wider roads are being planned on the basis that they, too, will relieve traffic congestion already stimulated by other new roads.

Their scale necessarily makes them appear to be far wider than roads really are. But the accompanying maps (figs. 13–16, plate 8) help to show the dense skein of roads already on the Blue Ridge and the degree to which the landscape has been fragmented.

The map showing roadless areas turns things inside out (fig. 17). On this map, "roadless" is defined by the Forest Service—and the words are worth

SOLUTIONS 11

Land Trusts in the Blue Ridge

Land trusts are nonprofit organizations that work through voluntary means to protect open land important to the quality of life and environmental health of their communities, states, or regions. Natural areas in the mountains are sometimes protected through voluntary easements, sometimes through donation or sale. Here is a partial list of land trusts in the Blue Ridge or national trusts with interest in the region:

Georgia

Mountain Conservation Trust of Georgia
104 North Main Street, Suite B-3
Jasper, GA 30143

Mountain Land Preservation and Restoration Society
712 Henry Drive
Clayton, GA 30143

Oconee River Land Trust
302 Riverview Road
Athens, GA 30606

North Carolina

Blue Ridge Parkway Foundation
P.O. Box 10427, Salem Station
Winston-Salem, NC 27103

Carolina Mountain Land Conservancy
P.O. Box 2822
Hendersonville, NC 28793-2822

Conservation Trust for North Carolina
P.O. Box 33333
Raleigh, NC 27636

Foothills Conservancy of North Carolina
P.O. Box 3023
Morganton, NC 28655-3023

Highlands Land Trust
P.O. Box 1703
Highlands, NC 28741

National Committee for the New River
P.O. Box 180
Glendale Springs, NC 28629-0180

North Carolina Botanical Garden Foundation
UNC Chapel Hill, Campus Box 3375
Chapel Hill, NC 27599

Riverlink
P.O. Box 15488
Asheville, NC 28813

The Southern Appalachian Highlands Conservancy
34 Wall Street, Suite 802
Asheville, NC 28801-2710

South Carolina   Naturaland Trust
P.O. Box 728
Greenville, SC 29602

Tennessee   Foothills Land Conservancy
352 High Street
Maryville, TN 37804-5835

Tennessee Conservation League
300 Orlando Avenue
Nashville, TN 37209

Tennessee Trails Association
P.O. Box 41446
Nashville, TN 37204-1446

Virginia   Potomac Appalachian Trail Club
118 Park Street SE
Vienna, VA 22180

Virginia Outdoors Foundation
203 Governor Street, Suite 316
Richmond, VA 23219

Western Virginia Land Trust
P.O. Box 18102
Roanoke, VA 24014

Regional   The Nature Conservancy
1815 North Lynn Street
Arlington, VA 22209
(or contact local field office)

Trust for Appalachian Trail Lands
Appalachian Trail Conference
P.O. Box 807
Harpers Ferry, WV 25425

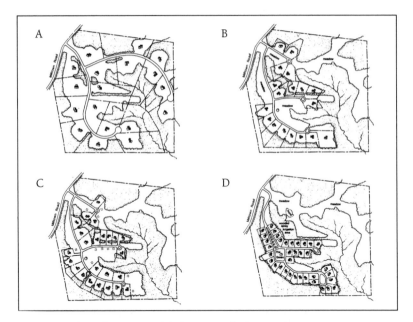

Figure 11. *Cluster housing.* Careful planning and "clustering" can preserve open space and natural habitat and enhance profit margins for developers. "A" shows a site zoned for eighteen lots in a traditional subdivision. In "B," there are still eighteen lots, but 50 percent of the open space is preserved, undivided. "C" shows twenty-four lots and 60 percent undivided space; "D" has thirty-six lots and 70 percent undivided space. (*Source:* Arendt, *Growing Greener,* 5-6)

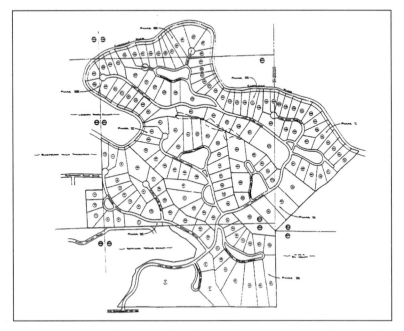

Figure 12. *A traditional subdivision along the Cartecay River in Georgia's Blue Ridge.* This use of land leaves almost no open space.

SOLUTIONS 12

Suburban Plan: Sustainability

[Virginia's Thomas Jefferson Planning District, which includes several counties along the Blue Ridge, created a Sustainability Council in 1994 to "describe a future where our economic, human, social, and environmental health are assured." A consensus of the thirty-four-member council of farmers, carpenters, builders, businesspeople, environmentalists, developers, and local and state elected officials produced the following principles. They are the basis for the council's detailed plan for reconciling future population growth and protection of the ecosystem.]

In a sustainable community:

- Individual rights are respected, and community responsibilities are recognized.
- All human and natural needs are respected, and conflict is resolved through consensus building. The community is a collection of diverse human and other biological interests.
- Achieving social, environmental, economic, and political health has intergenerational costs and benefits that must be weighed. In a healthy society, benefits outweigh costs.
- The integrity of the natural systems will be maintained or improved.
- Social, environmental, economic, and political systems are acknowledged to be interdependent at all levels.
- The responsibility for future generations' social, environmental, economic, and political health is acknowledged.
- The members understand there are limits to growth.

Source: Thomas Jefferson Planning District Sustainability Council, *Indicators of Sustainability.*

weighing carefully—as an area that includes no more than a half mile of improved road in each thousand acres. "Improved" means that some logging roads and other primitive roads are ignored in the calculation. Even employing this generous standard, "roadless" areas only total about 3 percent of the land area of the Southern Appalachians, and more than a third of that is within Great Smoky Mountains National Park.

The price tags hanging from new roads that steadily arrive in the mountains are ponderous (see Solutions 14). Georgia is mulling over a 180-mile-long, $476 million "Appalachian Scenic Parkway" that would span the northern part of the state from Trenton to Lavonia, near Toccoa. About 70 miles of this corridor would be four-lane highway. The route borders steep Blue Ridge foothills for 90 miles at the edge of the national forests, crossing

SOLUTIONS 13

Planning for Growth in Mountain Communities

Population growth management can be crucial for mountain ecosystems. If they choose to use them, Blue Ridge towns and counties have several tools for shaping future development to help protect natural areas. Wayne Strickland, Virginia's Fifth Planning District executive director, points out a few:

- Zoning, the dreaded "z word" in some jurisdictions, can protect private property owners, community values, and natural areas. Once a community decides how it wants to grow in the future and which land features need sensitive planning, zoning can help control future growth and avoid sprawl.

- Cluster development—developers can boost property values and achieve equal or even greater densities while protecting viewsheds, watersheds, forests, and wildlife habitat, with clustering (fig. 11). But first, local governments must move beyond traditional subdivision plans that resemble a collision of piano keyboards (fig. 12).

- Owners can agree to protect property from development with a conservation easement for a period of time or in perpetuity.

- The public can buy conservation easements and land with local, state, or federal funds. For example, North Carolina senator Jesse Helms sponsored an appropriation of $750,000 in federal funds in 1997 for the acquisition and conservation of private lands along the Blue Ridge Parkway.

the headwaters of the upper Chattahoochee, Chestatee, and Etowah Rivers and many smaller streams, as well as bear and other wildlife habitat.

The "preferred" route for the Andrews-to-Almond project in North Carolina, a four-lane highway through the Nantahala National Forest, intersects the Appalachian Trail in Stecoah Gap. The plan includes two tunnels that total 1.5 miles in length. It may become the most expensive, in average per-mile construction costs, ever built in the state if it is completed. The total estimated outlay for all these projects in North Carolina's portion of the Blue Ridge will be $1 billion—part of the state government's goal "to ensure that 96 percent of state residents will have a four-lane road within a 10-mile reach."

There is talk of a multimodal "Trans-America Highway" from Norfolk to Los Angeles, with a Blue Ridge crossing in the Roanoke-Bedford area. In Tennessee, Interstate 26 across the Unaka Mountains and the Cherokee National Forest between Erwin and Sam's Gap was completed in 1995. It is planned to connect with I-26 in North Carolina.

- Developers can proffer open space or other protection for natural areas in order to obtain advantageous rezoning.
- Land with higher development potential in nonsensitive areas can be traded for acreage in sensitive natural areas.
- "Rural communities have to pay, every time a house is placed on a lot," Strickland notes. Tax impact statements can be drawn up by local government agencies to analyze the true costs to taxpayers of new development: schools, social services, roads, water, sewer, and other demands on municipal infrastructure.
- "Some states such as Maryland are saying they can't afford leapfrog development anymore," Strickland says. "Smart growth" legislation there and elsewhere encourages new development in areas that can already provide infrastructure economically. It restricts growth in agricultural and natural areas, to protect them and to save the tax dollars that sprawl consumes.

As Strickland notes, however, "It's an issue not only of the tools you have available, but also the political will."

Source: Strickland interview; U.S. Congress, "Report 104-863," 1002 (for parkway funds figure).

A 16-mile section of the four-lane Foothills Parkway, which skirts the Smokies Park, was scheduled for completion in 1998. Another 35 miles is to be added in the future.

Virginia plans to widen parts of U.S. 221 to four lanes, including a section that comes within a quarter of a mile of the Blue Ridge Parkway. Parkway superintendent Gary Everhardt says such projects hasten the arrival of more traffic and "the slow creep of development."

"The eventual outcome could be the loss of the rural and forested mountain landscapes for which the Roanoke region is known," Everhardt wrote state highway authorities. "The impacts of improving a road go well beyond just the soil and vegetation that gets moved around in the construction limits of a project."

No one has come up with much research on alternatives to the road-building treadmill in mountain and rural areas, according to Bill Klein, director of research at the American Planning Association. The rapid-transit and ride-sharing schemes promoted in urban areas do not

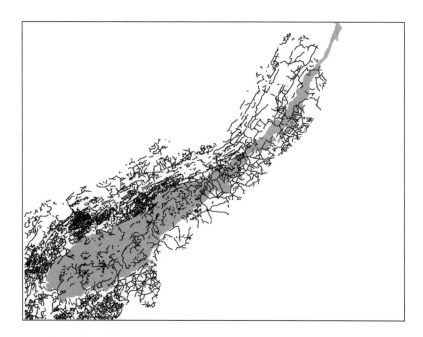

Figure 13. *Primary state and federal highways.* The least-roaded patch on this map and the ones following, a notable contrast with the rest of the Blue Ridge, is Great Smoky Mountains National Park. For a rough idea of the scale, remember that the Blue Ridge ecosystem is about 80 miles across at its widest point. (*Sources:* Keys et al., *Ecological Units of the Eastern United States*, CD-ROM; Hermann, *Southern Appalachian Assessment GIS Data Base*, CD-ROM)

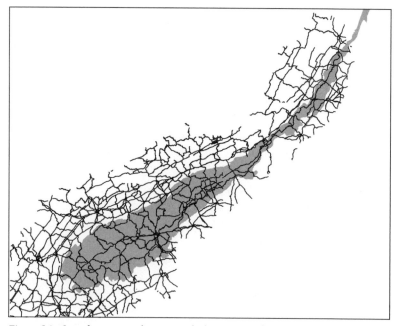

Figure 14. *Secondary state and county roads.* (*Sources:* See figure 13)

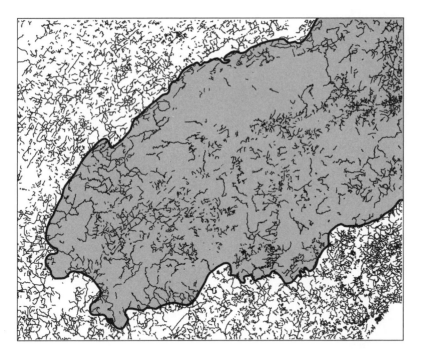

Figure 15. *"Light duty," local traffic, all-weather paved, and hard-surface roads.* Only the southern Blue Ridge is shown, for clarity. (*Sources*: See figure 13)

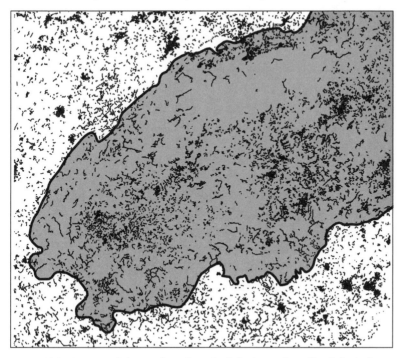

Figure 16. *Unimproved, dry-weather-only roads.* Only the southern Blue Ridge is shown, for clarity. (*Sources*: See figure 13)

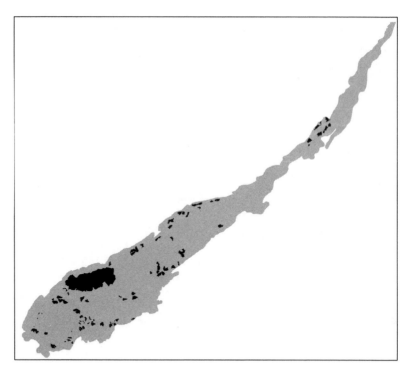

Figure 17. *"Roadless" areas in the Blue Ridge.* A Forest Service proposal to protect these small and rare, relatively unroaded areas from construction of additional roads provoked fierce debate. "Roadless" areas make up only 3 percent of land in all of the Southern Appalachians region. More than a third of it is within Great Smoky Mountains National Park. (*Sources*: Keys et al., *Ecological Units of the Eastern United States*, CD-ROM; Hermann, *Southern Appalachian Assessment GIS Data Base*, CD-ROM)

lend themselves to widely dispersed populations in places like the Blue Ridge.

"The solution is: don't do 'leapfrog' developments in those areas," he says. "Other than that, there are no real alternatives." "Building roads creates more traffic and more demand for more new roads. We've known that for a long, long time. But we can't seem to do anything about it, politically."

Economic issues overlap with concerns about what is efficient, or ugly, or likely to enhance our quality of life. The open, natural areas that mountain economies increasingly depend on can change quickly.

Everhardt and others argue that the transformation threatens not only the natural beauty of the parkway but the economic boost it brings to the

SOLUTIONS 14

Costs of Some New Roads in North Carolina's Blue Ridge

Widen 11.6 miles of U.S. 221, Blue Ridge Parkway to N.C. 181, to a multilane road: $40.2 million (identified future need)

Widen 14.5 miles of U.S. 221, Route 226 to the parkway, to a multilane highway: $39.3 million (construction to begin 1998)

Widen 15.3 miles of U.S. 321, from N.C. 268 to north of Blowing Rock, to a multilane highway: $74.9 million (construction begun 1997)

Widen 12 miles of U.S. 421, from N.C. 194 in Boone to 2 miles east of U.S. 221, to four-lane highway: $70.9 million (construction under way 1996)

Widen 13.6 miles of U.S. 421/321 from the Tennessee state line to U.S. 221 in Boone and reconstruct Blue Ridge Parkway Bridge over U.S. 421: $61.6 million (identified future need)

Widen and improve 30 miles of U.S. 19/23 to four-lane freeway, from I-240 in Asheville to Tennessee state line: $298.3 million (construction under way 1996)

Widen N.C. 60 from Georgia state line to U.S. 66/74, 4.8 miles: $15.4 million (construction under way 1996)

Construct two-lane road from U.S. 19 in Murphy to east of N.C. 141, ultimately to become four lanes, 4.8 miles: $47.4 million (construction to begin 1999)

Construct new four-lane U.S. 74 route from Andrews to N.C. 28 east of Almond, 27.1 miles: $341.4 million (partial construction under way 1996)

Source: North Carolina Department of Transportation.

mountains. "I think we're going to kill the golden goose if we aren't careful," he says.

"You have to be a nice place to live, in order to be a great place to visit," Virginia planner Katherine Imhoff says of the mountains. "I think our struggle has been and will be, increasingly, to retain the features that make us special, and a place people want to spend their money and time to come and see. Not Anyplace, U.S.A. I think the trend right now is toward Anyplace, U.S.A."

The economics and aesthetics of population growth concern just our own species. For the natural ecosystem, population growth—and the forms that our new constructions take on the land—will have consequences that reach much further.

SOLUTIONS 15

Jobs, Growth, and Seed Corn

[John M. DeGrove is director of the Florida Atlantic University/Florida International University Joint Center for Environmental and Urban Problems.]

Until recently we had a place in Hayesville, in Clay County, North Carolina. We sold it and moved to Sylva. There are virtually no rules there, either. They're cutting on steep slopes just over the ridge in back of me. At least there's a lot of argument about it these days: whether to do it, does it make sense, isn't there a better way to harvest this timber on a more sustainable basis?

I have worked over the years in what is called growth management, what I call "how to grow smart instead of dumb." It's difficult to convince people in the mountains that any restriction on land use could do anything other than drive them further into poverty—a bunch of Piedmont people trying to take jobs away.

Ironically, unless they change their approach, they're destined to gradually let a bunch of people from outside, including rape-ruin-and-run developers, mess things up, and it won't end up helping them.

It may create some kinds of service jobs, but not the decent-paying jobs folks really want in the mountains, so their children wouldn't have to leave.

I was born on a one-sick-mule farm, and I really get along well with my mountain neighbors. A lot of them are awfully good people. They understand to some extent that polluting the water and messing up the land isn't a good thing, but they can't bring themselves to make the leap from that to getting together to figure out how to do something about it.

Understand, I don't mean "stop the world I want to get off." If you protect the natural systems of the mountains, and protect the beauty of the mountains, if you don't skin off all the ridges and put ugly high-rises on them and foul the rivers, you maintain their attraction for quality development by quality companies, because their employees will think it's a good place to live.

"Sustainability" is the word you hear now. I think of it as "don't eat your seed corn." But if you don't have any land-use planning, no rules, no zoning, you take what comes.

Over the years, I've seen a steady decline in the mountains. Clear-cutting, stream pollution, sprawl development just scattered all over hell and back— no plans, no rhyme, no reason. We've been building a whole lot of roads without thinking very much about where we put them, or where the growth that they bring should go.

You can almost see the Appalachian Trail from parts of Hiawassee, Georgia. Hiawassee is a mess. You look up at the slopes as you ride through town, and they're still cheerfully skinning them off. I talked to a commissioner in a nearby county, who said, "They're just too damned greedy to stop and think what they're doing to themselves. We're not going to do that in our county." I hope he's right. A few developers in other places are maintaining a lot of open space, wide setbacks along streams, just commonsense things.

In fifty years, the Blue Ridge won't be a total disaster because there is a substantial amount of public land, and I assume we're not going to abandon all responsibility for protecting that and sell it off as some of my lawmaker friends have suggested we do. But I think many of the beautiful mountain vistas that nurture the soul—even for the people who live there and need jobs—are going to be gone, along with water quality and quality of life. A lot of places that are already going downhill are going to be a damned mess. Right now, we're eating our seed corn, right and left. We're growing dumb, instead of smart.

We need to come to some kind of limitation on population growth every- where, or else how the hell are we going to be sustainable? I do believe that. However successful we may be, good planning in the Blue Ridge is going to be cancelled out if we don't somehow bring population growth into balance with our resources. I'm an optimist, though. What's the alternative?

Source: DeGrove interview.

I n a natural landscape, one kind of ecosystem can shade into another in a gradual way. As you change elevations on a mountain slope, for example, the frequency of fog, the amount of rainfall, the temperatures and soils, and the mix of plant and animal species often shift slowly. Edges, on the other hand, are abrupt. They appear where the forest ends at a river, or a meadow stops at a rockslide.

The accelerating crisscross of new roads and other clearings on the Blue Ridge landscape proliferates edges and what biologists call "edge effects." In our eyes, these occur a few at a time, so they may seem inconsequential at first. For us, the illusion of intact forests persists well after they have been breached repeatedly. Other organisms, enmeshed within the web of the ecosystem, must respond to such changes immediately as they try to survive.

Roads that cross or are widened at, say, the Blue Ridge Parkway pose new barriers for animals, creating a string of habitat islands out of what were once long, unbroken stretches. To the degree that they are stranded, the animals cannot roam to find mates and avoid predators, food shortages, or other hazards.

The rapid accumulation of "edge" can be demonstrated with a pencil. Draw a square that is about 3 inches on a side, and imagine it as a forest. You have 12 inches of edge and an enclosed area of 9 square inches—an edge-to-area ratio of 4 linear inches to each 3 square inches.

Now draw an imaginary road down the middle, dividing the square into two rectangles. You have nearly the same total area, but 6 more inches of edge (a real road adds two new edges). The ratio of edge to area has increased to two to one. Add two more such roads, and the edge-to-area ratio is more than three to one. If we noninhabitants drive through, sealed in our car and moving pretty fast, we might not even sense the change. But as this imagined island of undisturbed forest becomes many smaller islands, the survival of the species that depend on big forests becomes increasingly marginal.

Some species—coyotes and raccoons, for example—can use these changes to their advantage. But for others, even at the level of salamanders and insects, foraging and breeding opportunities can increasingly be cut off as edges and their consequences multiply and intact areas divide and disappear. Life becomes incrementally more hazardous, and the relationships of species within an ecosystem change. Animals that have evolved within large forested areas are suddenly exposed to intense, sometimes overwhelming competition from species that thrive at forest edges or on human refuse.

The new 50-mile-long, $115 million Tellico Plains (Tennessee)–to–

Robbinsville (North Carolina) Highway, dubbed the "Cherohala Skyway," illustrates how edge changes the rules of the survival game. An environmental impact statement on the new road prepared for the Federal Highway Administration in 1977 stated that it would result in "minor disturbances" to animals because "the area has been logged in the past and is still being logged, and the animals and plants have become adjusted to the activities of man."

The final segment of the skyway opened in 1996. It crosses the Cherokee and Nantahala National Forests, what was once the Santeetlah Bear Refuge, the Tellico Bear Refuge, the Tellico Wildlife Management Area, the Falls Branch Scenic Area, perhaps a dozen trout streams, and what were patches of virgin hemlock and black cherry. All this was noted by the U.S. Interior Department in 1973, when plans were being reviewed. "This type of unique area (remote and isolated) is in short supply in the Eastern United States and construction through or adjacent to these areas should be avoided," the department stated then.

The Tennessee Fish and Game Commission, decrying "severe long-term effects on the mountain ecosystem," predicted siltation of trout streams and loss of habitat for bobcats and other carnivores, turkeys, raptors, amphibians, and reptiles. A road would "violate the concept of the bear refuge by the loss of habitat, isolation, remoteness, and food quality and will substitute human and vehicular activity which cannot be tolerated by the black bear," the commission's director stated.

Among other things, the federal environmental impact report missed the fact that the new highway bisects the habitat of what is now a federally listed endangered species, the northern flying squirrel. Biologists recommend that "where northern flying squirrels are found, the habitat should be vigorously protected from alteration, and managed to improve it for the squirrels."

The Department of Transportation called in Peter Weigl, a zoologist at Wake Forest University, to study the road's effects on the squirrel population in 1995 when construction was nearly complete. He had expected the animals to cross with relative ease. They could, after all, "fly" over much of the right-of-way, using the loose folds of skin between their front and back legs for gliding. And the road was not yet open to traffic.

The squirrels' movements were carefully monitored with radio collars and "mark and retrieve" techniques, and the findings of Weigl's research team were conclusive. "The road that they put in there is an absolute barrier," he said after several months of work. "No small animal crosses it."

The new construction walls off feeding and breeding options that were an aid to the survival of the small remaining population of squirrels. It strands them in both halves of the landscape. "Because you've cut into the side of a hill," Weigl says, "you have not just the width of the road, you have the removal of the vegetation on both sides, plus the boulder pile-up on the downhill side, which is very hard for an animal to traverse."

He found the tracks of coyotes and bobcats on the shoulders of the road, which they use for easy hunting. If you're a small animal attempting to avoid predators, the Skyway converts forest cover into a miles-long barren of pavement and gravel.

The new road is a lose-lose proposition for wilderness, too. Proponents argued that it would have little impact on the natural values of the area, though it parallels the Citico Creek Wilderness for several miles. It also runs just across a ridge from the Snowbird Creek watershed, which Congress requested the Forest Service to evaluate as a potential wilderness area. Snowbird was not recommended for wilderness status, however, because of its proximity to the new highway. "We thought the ease of access would eliminate the opportunity for solitude," assistant district ranger Frank Findley says.

Roadkill, a routine feature of rural life, is seldom tallied in such places. We may wonder about the collisions, though, as author Barry Lopez once did: "Who are these animals, their lights gone out? What journeys have fallen apart here?"

While some species of small mammals avoid roads, other populations prosper, even in the grasses along the interstate highways given the right conditions—though individuals are frequently killed. Their populations are relatively unaffected by roadkills if they are abundant, but many are not. Skunks, weasels, red foxes, gray foxes, minks, muskrats, bobcats, six kinds of turtles, and twenty-three kinds of snakes inhabit the Skyway area, for instance, often in smaller populations.

Traffic predictions for the Skyway vary widely, but the most recent federal estimate is 840,000 cars annually during the road's first years. On average, that's between 1 and 2 cars a minute, 24 hours a day, 365 days a year. After twenty years, the agency predicts, traffic will have nearly doubled.

There is a school of thought on the still growing human influence on natural areas that resolves such conflicts with ease. It requires that either the natural world conform to what we perceive as our own immediate needs or we are well rid of it. The agenda of what is called the "wise use" movement, for example, advocates curtailed legal protection for "non-adaptive" endangered plant and animal species. And radio entertainer Rush Limbaugh writes that the way things ought to be is, if a disappearing species "can't adapt to the superiority of humans, screw it."

American black bears are nothing if not "adaptable." So much so that they may test our notions of how much habitat we can withdraw and still expect wildlife to survive. Bears devour acorns and even eat rock-hard hickory nuts. They ignore the stings of yellowjackets while dining on nests packed with larvae. Bits of metal and glass and a 5-foot-long piece of plastic strapping have been found in the stomachs of Blue Ridge bears. Males sometimes eat their children. A female can den with a couple of cubs high in the too-small hollow of a poplar, 100 feet above ground, with snow falling on her rump all winter.

Bears can live in the steepest and most isolated terrain, where food is often scarce, because 90 percent of their former habitat in the southeastern United States has been taken for human use. Their kind survived the disappearance of a primary food source, the chestnut, until the oak forest matured to yield acorns.

They are hunted by teams of predators using radio-equipped, four-wheel-drive vehicles, high-powered rifles, and relays of radio-collared dogs. Mountain populations in the Southeast are stable or even increasing slightly, biologists say. The estimated bear count in the Southern Appalachians is about nine thousand. That, surely, is "adaptable."

But bear habitat outside public lands is shrinking. "All along the Smokies park, where they don't have a national forest buffer, you're getting a tremendous influx of people as tourists and permanent residents," University of Tennessee bear biologist Michael Pelton says. "Second home developments are going in, and leaving these animals fewer and fewer opportunities for their foraging activities at lower elevations—usually in the fall, looking for acorns.

"So that interface between where the park ends and the humans begin is becoming more and more well-defined. You begin to get more and more nuisance complaints and road kills, and the whole situation is aggravated because of the increasing human population."

As Pelton and a colleague have written, human population growth will be perhaps the most important influence on bear populations in the Southern Appalachians: "The sad truth is that the continued urban and suburban growth of the region probably precludes the existence of bears in those areas. . . . Without some level of management in the future the risk of crossing unsuitable habitats will likely become formidable."

Bears and people do not keep easy company. The "teddy bear syndrome" tempts us to approach bears, but food, not fellowship, attracts them to us. Though it is illegal to feed them in the national parks, in four hundred cases in the Smokies in which bears approached humans over a twenty-year period, food was first offered to the bears. When people or property suffer damage, teddy bears become "problem bears." They are trapped and shipped off to distant habitat where, researchers have found, the bears probably have less than an even chance for short-term survival.

When the acorn crop fails, as it does every few years, bears must search widely for food or they starve. The fall of 1992, for example, saw a poor acorn yield, preceded by a poor berry crop. As the bears moved out of their home territories to forage, many were killed by passing cars.

Roads revise the odds on survival in several ways. Bears will use forest roads—especially those that have been closed and gated—as travel lanes and feeding areas for berries in the summer unless traffic picks up. But roads give the high-tech hunters, including poachers, easy access to more bear habitat. These factors force the bears to shift their home ranges to less desirable areas, where starvation and other hazards are greater.

As development arrives with the roads, the animal has even fewer options for how, where, when to cross, and if to cross. "It's kind of a risk-reward deal, and in many instances the risk is getting too great," Pelton says. "That's a concern not just for bears, but for a lot of other species."

Females are especially road-shy. One study of bear behavior predicts that even in a 20-square-mile area, one improved road and a campground would reduce bear usage by more than half.

Interstate 40 between Asheville and Knoxville roars right past the Harmon Den Bear Sanctuary and Pisgah National Forest, walling bears off from Great Smoky Mountains National Park's bear population on the other side. Like the Cherohala Skyway and most other U.S. roads, I-40 was constructed by "cut and fill"—cutting into the sides of mountains and filling in the ravines rather than bridging and tunneling, as is often done in Europe.

By contrast, the Linn Cove Viaduct, which carries a curving section of the Blue Ridge Parkway around a sheer slope near Grandfather Mountain, has left the landscape comparatively unaltered and accessible for wildlife, as have the many tunnels on the parkway.

I-40, however, makes a safe crossing for bears and other animals nearly impossible. One tunnel creates a single small land bridge in more than 20 miles of mountain interstate highway. Research cameras have documented that at least a few bobcats, raccoons, and opossums use the drainage culverts to pass under I-40, along with domestic dogs, cats, and Norway rats. No bear was seen in the photos.

Instead, bears chance the "ribbon of risk" or "the moving wall," as roads are called in the research, when they must cross. "A 70- to 90-kilogram bear was observed around 8:30 A.M. on 29 December, 1995, on the eastbound side of Interstate 40 near the Harmon Den exit in North Carolina," one eyewitness report states. "The bear was wet from swimming across the

Pigeon River. It ran across the road and jumped on top of the concrete median and nervously watched several passing vehicles going west. The bear then crossed in front of oncoming traffic."

On the evening of July 27, 1995, University of Georgia biologist Byron Freeman was camped with his crew near the Dawson Forest, at the confluence of the Etowah River and Amicalola Creek, taking readings on the conditions of fish, the insects they feed on, and water quality.

Freeman knows the Etowah and its dozens of tributaries well after years of driving, canoeing, and bushwhacking its 165-mile length. He and his colleagues have discovered several hitherto unknown species of fish in this and other Blue Ridge streams.

One index of water quality is simply how great a load of dirt the river is carrying downstream. That is gauged by an instrument that passes a beam of light through a water sample to see how much of the light is deflected by particles suspended in the water.

In scientific parlance, this is "turbidity," the opposite of "clarity." Pristine mountain streams read out at about 3 on the turbidity index. On this evening, the Etowah showed a relatively clear 20.

"We woke up the next morning and the river was at 143, and we flipped," Freeman recalls. "It had been so clear the day before. We kept waiting for a major flood—we had heard thunder farther up in the mountains the night before—but it never arrived."

The "new" fishes Freeman and his colleagues have recently discovered in these drainages have probably emerged as separate species only in the last ten thousand years or so. Some have evolved into shimmering flags of courtship, as colorful as if they were misplaced from somewhere in the Caribbean.

One little perch, unnamed as yet, bears eight vertical green bars over orange patches on its sides; a sea-green rear fin with a broad red-orange band running through it; a green dorsal fin with blue-green marginal bands; two bright orange spots at the base of the tail; a series of rusty brown dots high on its sides; deep green pelvic fins.

These fish lay their eggs singly on the tops of cobble—rocks the size of your fist—and they require a clean spot. If the riverbed smothers under a blanket of sediment, they have no place to spawn.

Six hours of working upstream, taking readings below various tributaries, finally led Freeman and his crew to a site, miles distant, that proved to be the source of the pollution. There, at the head of a small creek, a new golf course was under construction.

"This is a big watershed, so if you look at a map, you just say holy cow, here's this little bitty place that generated all this mess," Freeman says. The pollution continued for several months. The lesson: even a tiny creek can pulse chronic, choking loads of silt into the ecosystem—enough to pollute a whole river miles away.

The headwaters of the Etowah emerge from the Chattahoochee National Forest in Lumpkin County, Georgia, near the county Route 72 bridge. It is perhaps 30 feet wide here, and you can see the submerged stones and boulders of any Blue Ridge upland stream. Except that they're embedded in a thick gruel of sand and sediment that once was topsoil. It has been flushed down out of the forest, fish biologist Noel Burkhead says, from logging operations on the national forests or on private lands higher in the watershed. "Best management practices" that are supposed to guide logging on private land are seldom checked on by state authorities, he says—the personnel and the political will are missing.

"I've seen private timbering sites that look like the side of Mount Saint Helens after the logging company is done," Burkhead says. "The landscape was totally trashed, just a complete mess, leaving acres and acres of exposed soil available to erode into streams."

Many kinds of fish in the upper Etowah make their lives on the stream bottom, spawning, feeding, and sheltering there. Within this chill world, all but invisible to us, commonplace but essential natural processes support life.

Consider, for example, the uses of the energy stored in a leaf that falls into the river. Lodged in the gravels and cobbles of the bottom, it is shredded by insects and bacteria. They are themselves food for fish, the fish are prey for larger species, and so on up the food chain.

But when the crevices and rock facets where life thrives are clogged with sediment, the stream is biologically sterilized, except for a few tolerant organisms. A complex web of life that recycles energy almost miraculously is then reduced to a surviving strand or two.

The Etowah is rich in biodiversity, supporting ninety-one native species of fish, including five found nowhere else in the world. But as you travel farther downstream, the waters cloud and diversity narrows. Eel and sturgeon were found in the Etowah until midcentury. They have disappeared, along with a dozen other species. "Some of these are hardy animals, like the chain pickerel, which is common," Burkhead says. "What do you have to do to an ecosystem to eliminate chain pickerels? Obviously, some pretty nasty insults, and in fact, the Etowah has had some."

This is not an exceptional, hard-luck watershed. Many Blue Ridge rivers and streams are gradually being suffocated by silt from slipshod development, logging, and road construction. Weak laws and lax enforcement accelerate the erosion, Burkhead says. Streams in north Georgia's Blue Ridge are in mediocre to poor condition and declining, he adds.

The gravels of Settingdown Creek and its tributaries, at the southern edge of the Blue Ridge, are typical of the advanced stages of the disease. They are packed with sediment.

During his research forays, Byron Freeman has also seen silt polluting the Chestatee, Toccoa, Little Tennessee, lower Tallulah, Ellijay, Cartecay, and Conasauga Rivers and the west fork of the Chattooga. One federal study of three hundred stream sites in the Blue Ridge and the "ridge and valley" region to the west found that 65 percent showed "moderate to severe degradation" from erosion and other human disturbances.

"The streams are being hammered with full-scale urbanization," Freeman says. "We are losing these systems before we even have a chance to discover what's in them. This is the rainforest in terms of biodiversity for temperate fishes, right here in the Southeast, and it's at risk if we continue to develop the landscape in the same fashion that we developed the Piedmont."

Of all the creatures that might suffer from edge effects, birds—airborne, even more mobile than we are—might seem to have the fewest problems with edges. Some, indeed, are doing fine. But surveys show that in the Blue Ridge, the populations of seventy-six species of birds have declined since the mid-1960s. Sixty-seven of those have declined by 28 percent or more. In the most cautious reading of the statistics, thirteen species are most distinctly in trouble (see Solutions 16).

Bird populations vary continuously with weather events, food availability, and many other natural factors, so researchers have been slow to conclude that human influences are at work. But as each year's statistics from thousands of systematic Breeding Bird Survey (BBS) reports have been analyzed since 1966, population trends have emerged more clearly. Fish and Wildlife Service biologist William C. Hunter, on loan to an international migratory bird conservation program, is among those who analyze the numbers.

"The striking thing about the Blue Ridge data specifically is the number of species showing declines," Hunter says. "The proportion of these species seems to be much higher in the Blue Ridge than in most of the other areas. My feeling is what we're seeing in the Blue Ridge is the result of second-home development, and the associated spread of stores and support areas and roads."

The rates may seem small, but this clock ticks loudly: the estimate of a 1.1 percent annual decline trend among ovenbirds in the Blue Ridge, for example, means that their populations have declined by more than a fourth during the history of the survey. No extinctions are currently predicted—only the conversion of several species of songbirds from "familiar" to "seldom seen."

The idea of extinction is important. It is, among other things, the final cancellation of the possibility of recovery. But chain reactions in an ecosystem, often unknown to us until they are well under way, do not wait for extinctions. Declines among bird populations can, for example, lead to an explosion in the numbers of insects they fed on. Insects, in turn, can have a direct effect on the health of forests.

Such relationships may have seemed conjectural only a few years ago—a way of stretching the point that every species has some vital role to play in the ecosystem. Then two midwestern biologists designed an experiment to test the idea. They artificially subtracted wood thrushes, warblers, tanagers, grosbeaks, vireos, and all other birds from an ecosystem inhabited by thirty young white oak trees. They enclosed the trees in net cages to keep the birds out, and an equal number of trees of the same size were left uncaged.

For several months, the researchers counted the insects on each of the oaks and carefully evaluated leaf damage. The role of the birds was far larger than suspected. "Birdless" oaks had twice the number of insects on their leaves and lost twice as much leaf matter.

The effects accumulate. The next year, the birdless trees were sicker, producing about one-third less leaf matter. The researchers concluded that birds protect young trees and help them grow by eating leaf-chewing insects. As bird populations decline, the trees decline. Birds help keep trees healthy. Over longer periods, fewer birds may even lead to changes in the mix of tree species in the forest.

Their study "probably underestimates the impact of insectivorous birds on plant growth," the researchers added, because the populations of three kinds of insect-eating birds had already declined there over the preceding decade. Action to maximize bird species diversity and maintain their populations is "imperative to maintain forest productivity."

Scientists differ on which factor is most important in explaining why populations of migrant birds in places like the Blue Ridge are dropping. Clear-cutting in U.S. forests, human population growth, and the destruction of tropical rainforests on the birds' wintering grounds in Latin America have been studied, along with the huge market south of the border for outlawed U.S. agricultural pesticides.

But a series of recent studies has established that wherever it happens, the fission of large, undisturbed forests into a landscape of roaded woodlots can overwhelm migrant songbird populations. Hunter sums up the

SOLUTIONS 16

Disappearing Songbirds

The Breeding Bird Survey involves thousands of trained volunteers. It has two strong virtues that bolster its credibility: the sheer volume of data it generates, and the fact that it spans three decades. Even when subjected to various statistical corrective measures, however, BBS results must be viewed cautiously, especially when single species or smaller regions such as the Blue Ridge are under discussion.

A process of elimination, in pursuit of increasing certainty, begins with the fact that ninety-nine Blue Ridge bird species have been monitored by the survey since 1966. The populations of seventy-six of those species declined during that period—sixty-seven of them by 28 percent or more.

Only fifty-six species showed up on enough survey routes to consider the data reliable, BBS coordinator Bruce Peterjohn says. Of that group, the declines for twenty-three species are statistically significant.

Even if the decline trend is in the range of 4 to 5 percent per year or greater, it should still be viewed with caution, Peterjohn says, so let's eliminate everything showing less than a 4.3 percent annual decline. That leaves a list of thirteen species in trouble.

At this point in the winnowing, we can be confident that the downward plunge in bird populations is real, "but I would have little confidence in the estimated magnitude of the rate of population change," Peterjohn says. With our pared-down list in hand, we know, too, that we may be ignoring some species that are declining steeply, although they didn't meet the criteria.

research: "These birds just get walloped by cowbirds and predators in small forest patches." The predators include raccoons, skunks, opossums, house cats, chipmunks, snakes, and egg-eating birds like American crows and blue jays. They all prey on nests far more frequently at forest edges, and raccoon populations swell where garbage is available.

The cowbird is also an edge dweller, feeding in open areas of short grass or bare ground. When a forest is fragmented by pastures, roads, building lots, or even mowed grass, cowbirds can feed, so they invade the remaining patches of trees. Cowbirds are parasites that locate the nests of other birds and destroy their eggs. Then the cowbird lays its own eggs, to be hatched and fed by the unwitting foster parents. Cowbird chicks are aggressive and grow quickly, so they can out-compete the nestlings of smaller species such as warblers, which often starve.

One experiment in the Smokies was designed to test how well birds fare in smaller sections of woods as compared with large, unbroken forests.

Declining Blue Ridge Bird Populations

Species	Estimated Average Annual Decline Rate (rounded to nearest 0.1%)	Estimated Decline from 1966 to 1996 (rounded to nearest 1%)
Black-and-white warbler	−7.5	−90
Eastern wood-pewee	−7.0	−89
Yellow-breasted chat	−6.7	−88
Kentucky warbler	−6.4	−86
Eastern meadowlark	−6.2	−85
Northern bobwhite	−6.0	−84
Field sparrow	−6.0	−84
Common yellowthroat	−5.6	−82
Chestnut-sided warbler	−5.6	−82
Chipping sparrow	−5.1	−79
Ruby-throated hummingbird	−4.6	−76
Barn swallow	−4.3	−73
Wood thrush	−4.3	−73

Sources: Peterjohn interview; Sauer et al., Breeding Bird Survey.

Quail eggs were placed in artificial nests in wooded areas of varying sizes in both rural and suburban areas, and in the broad forests of the national park. Almost 100 percent of the eggs were destroyed by predators in some of the smaller patches. In the national park, the predators raided only one nest in fifty.

Changes that seem modest to us can jar the natural landscape. In the Cherokee National Forest, for example, about 30 miles south of Knoxville, a series of power line easements and transmission towers sweep up the western slopes, cross Highway 129, and march down the other side. One-lane dirt roads wind over the same ridges under a canopy of chestnut oaks. How do migrant birds and their natural enemies respond to these three different kinds of openings and edges—power line rights-of-way, dirt roads, and two-lane highways—in what was once a large expanse of forest?

During one recent spring and summer, researchers tried to sort it out. They found that migrant songbirds that inhabit broad, unbroken forests tend to avoid the paved roads and power-line clearings, sensing correctly that these wider openings offer greater threats. But the narrow dirt roads with infrequent traffic are a different story. They create "ecological traps" for the migrants. These edges attract plenty of cowbirds and nest predators like raccoons and skunks. But unfortunately, and to their hazard, the migrant songbirds take little notice of and don't bother to avoid such narrow openings. They are tricked into becoming even easier prey.

"We suggest that these widespread corridors may be inconspicuous but important contributors to declines of forest-interior nesting species in eastern North America," the researchers say.

Bird population specialist Robert Askins says such results show that conservation efforts must be regionwide: "No matter how carefully they are protected, small nature preserves may progressively lose their most distinctive species if they are surrounded by a hostile landscape," he writes.

Meanwhile, no one can really paint a clear picture of bird population trends in the United States. The image only resolves itself into a mosaic of fuzzy blobs, assembled from shifting flocks of statistics, subject to varying interpretations. Florida State University ornithologist Frances James and colleagues reanalyzed Breeding Bird Survey data and concluded, for instance, that migrant bird populations across the United States are not generally dropping after all. But even she has called the Blue Ridge a "hotspot" because of the consistency of population declines over a large number of species during the thirty-year span of the BBS.

"That was very surprising to us," she says. "Even for species that are doing well elsewhere, sometimes we see declines in the Blue Ridge." Though the data are at times spotty, she says, "I think the conclusion is valid that there is a particular problem in the Blue Ridge."

Nothing in nature stays the same. New species may join the "in trouble" list in the future. Some species may disappear even more rapidly, some may plateau, and some may increase in number. "In reality, few rates of population change are consistent for 10 to 20 years, let alone for longer periods," BBS director Bruce Peterjohn points out. But if we assume that the current rates of decline will continue, the populations of nine Blue Ridge bird species will have dropped by 99 percent or more by the year 2050; another four will have declined by 97.5 percent or more.

"We in government cannot formally push the data forward, just because our government supervisors tend to get very nervous when we start playing those games," former BBS coordinator Sam Droege says. But forecasting in this way is legitimate, he adds, as long as it is clearly understood that the numbers convey only strongly disturbing general indications, not predictions. Cities and intensively farmed areas are devoid of all but a handful of bird species. Suburbs are "only marginally better," Droege says. This is especially worrisome on migrants' breeding grounds in North America. "We stand better chances of extinctions now because we're dealing less and less with fluctuations and more and more with unidirectional change," he says. "The population pressures are so high in certain areas that landforms and land types and land uses are getting locked in. Urban areas don't go back to being un-urban areas."

The Sierra de Los Tuxtlas in southeastern Mexico is not much like the Blue Ridge. Two 5,000-foot volcanoes and the northernmost tropical rainforest on the continent are its most prominent features.

But the two mountain ranges do have some things in common, including the wood thrush. It is not unlikely that some of the same individual thrushes tending their nests in Shenandoah or the Smokies in the spring migrate back each winter of their brief lives to Los Tuxtlas.

The wood thrush inhabits rainforests from Mexico to Panama during the winter season. Its spring and summer breeding range extends roughly across the eastern half of the United States. Robin-sized, feathered mostly in shades of brown, this is not the most eye-catching of migrants, but its song is among the sweetest. "Whenever a man hears it, he is young, and nature is in her spring," Thoreau wrote. "Wherever he hears it, it is a new world and a free country, and the gates of heaven are not shut against him."

Though it is still a fairly common bird, wood thrush populations are falling fast—about 30 percent in just ten years, according to the Breeding Bird Survey. It is especially susceptible to cowbird infestations. But the increasing problems on its breeding range in the United States do not completely explain the wood thrush's ill fortune.

"It's interesting, because it is showing declines just about everywhere," Chuck Hunter says. "That is unusual, because most of the other birds we're looking at have net losses when you add it all together, but they show declines in some areas and increases in others."

That the wood thrush is fading even from North American forests that are not split up into smaller patches suggests strongly that it is also under pressure on its winter habitat, thousands of miles to the south.

Smithsonian biologist John Rappole sees wood thrushes in nesting season in the Blue Ridge and during his winter research at Los Tuxtlas, where, in the 1950s, about half of the bird's original rainforest habitat remained. Satellite imagery showed that, thirty years later, all but a fraction had vanished.

"Immediate action is urgently needed to protect these small but rich remnants and preserve them," a biologist at the National University of Mexico has warned.

The nobly named Los Tuxtlas Mountains Biosphere Preserve is only a preserve on paper. Population pressure and economic interests have driven logging, subsistence farming, and cattle ranching onto higher, steeper slopes. "You're talking about slopes of 45 degrees, rainfall of 160 inches a year," Rappole says. "They get two years of use out of a field. There isn't much forest left. At some point they're going to run out, and they're still going to be facing the same problems.

"We started work there in 1973," he adds. "Since that time the amount of forest that was left has declined by half, replaced by pasture. There are no wood thrushes in pastures." Rappole has estimated that 95 percent of wood thrush habitat in Los Tuxtlas is gone, and the bird has nearly disappeared from the remaining area.

The wood thrush is not the only forest-dependent bird in these mountains. Seventeen other such species commonly winter there, and another thirty-three pass through. "Nor is the intensity of forest loss in the Tuxtlas unique," Rappole has written. "Many other parts of Middle America have undergone similar habitat alteration."

Despite these brightly colored, lighter-than-air links between them, the Blue Ridge and Mexico may appear to us to be remote from each other. Mexico seems a place apart, a Third World. In Mexico, they have a population problem. People there consume dwindling and irreplaceable natural areas as their numbers grow.

We rarely think of our own country that way, though the U.S. population has more than doubled since 1940. But it's likely that, in the eye of a wood thrush flying over, the landscapes of Los Tuxtlas and parts of the Blue Ridge have much in common.

No one really knows what the forests of the Blue Ridge would be like without our influence, though scientists are trying to puzzle it out. We have altered the nature of the mountains so frequently that the qualities of a "natural" forest are speculative, like poetry reconstructed from a lost language, or the sound of some ancient music.

How often did lightning fire the trees, and how intensely did they burn? How frequent were changes wrought by droughts, ice storms, hurricanes, and floods, and how do forests respond? Under natural conditions, how do plants and insects and animal populations influence each other? Our strongest clues to those natural harmonies and cross-rhythms lie in the small and scattered remnants of the forests we call original, virgin, primary, uncut, or "old growth."

It's not apparent how pristine even those venerable stands might be. Native Americans repeatedly burned extensive tracts in the mountains to promote game populations and agriculture. Historical geographer Michael Williams writes that they were "a potent, if not crucial ecological factor in the distribution and composition of the forest."

Tiny charcoal grains deposited over eons in a bog in Horse Cove, near Highlands, North Carolina, substantiate this view, at least in part. The researchers who analyzed these traces of past forest fires found that "during most of the last 4,000 years Native Americans played an important role in determining the composition of southern Appalachian vegetation through selective use of fire."

In some mountain forests, the mark of humankind in pre-European times may have been far less distinct. The study, by Hazel and Paul Delcourt of the University of Tennessee, also found that human-caused fires were concentrated on ridges and bottomlands along major rivers. Some areas along the steep eastern Blue Ridge escarpment and on north-facing slopes probably escaped such fires.

Four hundred years ago the era of European settlement began. Since then, nearly all of the Blue Ridge has been cleared, sometimes repeatedly. The most thorough erasure of forest ecosystems occurred between about 1890 and 1930, when industrial logging and subsistence farming devastated forests, soils, and streams over millions of acres of even the steepest terrain. Estimates vary in detail, but they concur on the general picture:

- By the turn of the century, 80 percent of the forestlands of the central and southern Blue Ridge had been been burned over, often several times.
- Despite sometimes thin soils and steep slopes, one-fourth had been completely cleared for cultivation, by fire or other means.

- A U.S. Senate committee report of 1908 found that "the Appalachian region has suffered incalculable damage from fire, which in many localities still burns every year unchecked."

All but a tiny fraction of the remaining original forest, and much of the regrowth, would be logged and burned during the thirty years that followed.

The significance of what remains of the old-growth forest may not be immediately apparent. To some, woods are woods, whether young or old. Past a certain age, after all, a tree's health declines, and insects and disease begin to degrade salable wood. Mature trees add wood more and more slowly, so they are not, in economic terms, productive. In order to grow more wood, you have to cut the wood that's already there. And endless rows of lab-bred "supertree" pines laid out in our national forests or on a paper company plantation make a fine sight—an abundance of tall spires that give every appearance of health, squared off like a great lawn.

A lawn's natural history is short, of course, repeatedly cut flat and reset at something near zero. Its chief virtue is uniformity and its music, a careful monotone. Like grass, plantations of trees grown as crop wood are often single species, even-aged, thinned, fertilized, treated with insecticides and herbicides, and tamed right down to their genes. Manipulated for maximum timber growth in the shortest time, they are a success story in the annals of modern agriculture.

But a farm is not a forest. Far fewer kinds of life, from microorganisms to mammals, subsist on these synthesized "green desert" landscapes. By definition, nature recedes as domestication intensifies—that's part of the point of agriculture.

Old-growth forests on the other hand are wild, complex, nonuniform, wasteful on the ledgers of short-term profit, but rich in the larger context of biological diversity. They include many species and ages of trees and support a broad diversity of organisms in an intricate pattern of relationships that evolve over centuries.

Blue Ridge forestlands can be found at many different points on the wild-versus-farmed spectrum, reflecting their history and the varied philosophies under which they have been managed. But "old growth"—that is, nearly untouched forest—is exceedingly, and increasingly, rare.

Only rough estimates have been made so far, but depending on how you define old growth—never cut; selectively logged in the distant past; or more than a century old with little visible disturbance—less than 1 percent to 3 percent of it may remain in all of the Southern Appalachians.

The Forest Service has compiled an inventory of possible old growth in the national forests but acknowledges that it is frequently inaccurate and incomplete, merely a beginning step. It overlooked, for example, some recently verified old-growth stands. In some places, a more accurate inventory is under way.

Patches of never-logged forest still turn up on the Blue Ridge, but they are well hidden. "Chances are, anyplace you can get to easily is not old growth," says ecologist Tom Rawinski of Virginia's Department of Conservation and Recreation. He points to the tightly compressed elevation contours on a topographic map of the James River Face Wilderness. There, along an unnamed ridge between Sulfur Spring Hollow and the Devil's Marbleyard, he recently came across perhaps a hundred acres of old growth.

Seeing it means an hour of bushwhacking through a steep fretwork of fallen logs and in-your-face catbriar and up a long bright expanse of quartzite talus studded with gnarled beeches and pines. "Back then, if they were doing horse- or ox-logging, they tended to avoid bouldery terrain," Rawinski says, working his way upslope. "You could really mess up your animals that way."

A dense overhead canopy of hemlock, chestnut oak, and tulip poplar branches shades the lower elevations. It is ripped open in places, probably by the passage of Hurricane Hugo in 1989. Some of the trees are at least a couple of hundred years old; some are only saplings.

Blowdowns admit light, Rawinski explains, which stimulates a burst of new growth on the forest floor. They also yank up shaggy, clodded Medusa-heads of roots and open big holes in the soil. This "pit and mound topography," characteristic of old-growth forests, makes new places for small mammals, insects, and herbs to live. The fallen trunks rot in place over several decades, providing more habitat and nutrients for the thin soils.

When one section of forest is leveled, whether by chain saws or insects or an ice storm, new trees will spring up in a season or two, but not necessarily the same mix of species. The types that will grow fastest and compete successfully will be those most tolerant of sunlight. They will create an upper canopy, or overstory, as they grow.

The shade of the canopy fosters a new generation and a different mix of species. They are more able to survive low-light conditions, and they grow more slowly. The generations of trees that follow will be made up of ever more shade-tolerant species. These stages are called "succession."

When roughly the same mix of species is coming up to replace the mature overstory, this patch of forest is said to have reached a "climax" successional stage—a relatively stable assemblage that awaits the next natural disturbance. Reaching that stage can take four centuries or longer.

Rawinski pauses along a steep and stony ridge. There, some of the natural history of this place is preserved in tree rings, visible in the hollow of a huge old chestnut oak. Wet and dry years are recorded in the varying thicknesses of the rings, and a thin, charred stratum shows under a century of newer wood. A low-intensity fire crept along the forest floor in that long-ago season, clearing the underbrush but sparing the trees.

"What has this plot of land gone through, and what would be lost if you cut all the trees down?" Rawinski asks. "You'd lose any evidence to piece together past climates, for one thing, or how often fires came."

Knowledge of natural forest conditions is crucial to restore ecosystems to health—or even to define what "health" is—and we still know quite little. For example, scientists have great trouble measuring the effects of air pollution, tree diseases, and climate change on Blue Ridge forests. Farming, logging, roading, erosion, replanting, herbicides, and other human activity can form a nearly impenetrable screen of aftereffects that confound the data and make single factors hard to discern.

The role of fire in some ecosystems presents one such puzzle. Several biological communities have evolved with and adapted to a natural sequence of lightning fires. But we have suppressed forest fires in the Blue Ridge since at least the 1930s.

Scientists warn that low-intensity prescribed burns—that is, fires purposely set and controlled—are needed immediately to maintain or restore several of the rarer communities, including high-elevation balds, mountain bogs, and birch stands in boulder fields.

"In the absence of periodic fire, two of the five rare forested communities . . . mountain longleaf pine woodlands and Table Mountain pine/pitch pine woodlands, are being replaced by hardwoods and loblolly pine," according to federal research. Even common ecological communities such as pine and oak forests may be threatened in some places because we have been so successful in eliminating fire.

But how often is burning needed in each of the many types of forest? The natural history recorded in old trees and relatively undisturbed ecosystems can sometimes tell us. "Old growth forests serve as controls for understanding the impacts of forest management practices," Middlebury College biologist Stephen C. Trombulak has written. "As the amount of old growth declines, our ability to assess scientifically the appropriateness of ecosystem management . . . or any other management philosophy, declines."

A few miles south of the James River Face Wilderness, Rawinski and his colleagues have located another old-growth forest of unknown extent, but probably several hundred acres, near the Appalachian Trail on Apple Orchard Mountain.

Along a boggy flat, thick with rhododendron, the shattered trunks of dead hemlocks 4 feet in diameter form a kind of bear motel. The coral-colored wood is clawed and wallowed into big soft mats of fine splinters, fit for a bear to lounge on in secluded splendor.

Many such species thrive in old-growth forests. Diversity offers survival options; uniformity limits them. Especially where they are heavily hunted, bears also like to den in the hollows of old trees. The extreme scarcity of old growth narrows their choices.

Other carnivores, too, such as foxes, minks, otters, weasels, and bobcats, typically shy and secretive, seek out deep seclusion and secure refuges. "Their escape cover, maternity sites, and winter dens are provided by components of older forests," biologist Michael Pelton of the University of Tennessee writes. "Because of the history of logging in the East, it is clear we should preserve all remaining old growth and create more."

A forest composed exclusively of old trees would be as unnatural as any other kind of uniformity. Younger patches provide more browse and berries for deer and bear and a different assortment of food and cover for a variety of birds, insects, and small mammals adapted to new growth.

These admixtures of younger and older forest are created continuously by natural disturbances. An old-growth forest is not a museum but a dynamo, constantly re-creating itself. Fires, hurricanes, ice storms, and other disturbances are an essential part of that self-renewal. They are sometimes wryly called "nature's clear-cuts" by clear-cut logging advocates, but that argument blurs sharp differences. Young forests developing after natural catastrophes such as wildfires or hurricanes have large standing dead trees and fallen logs and a nearly undisturbed forest floor. Dead and dying wood also yields organic matter and nutrients for the soil. As a Forest Service report notes, "Dead, standing trees can provide tree cavities that are important to many animals. When snags fall, they become logs on the forest floor or in streams that provide microhabitats for a diverse succession of organisms."

Logging of any kind occurs in addition to, and multiplies the effects of, the cycles of natural disturbance. In the more moist, old-growth hardwood cove forests in the Smokies, natural disturbances open about 1 percent of the canopy, on average, each year. In other Blue Ridge forests, more affected by weather and fire, gaps in the overhead canopy occur more frequently. Unlike clear-cuts, low-intensity natural disturbances result in forests that are physically complex and uneven in age.

The new growth that follows logging can provide a variety of habitats and enhance biodiversity, too. That kind of variety, however, is already well represented on the three-fourths of Blue Ridge timberlands that are privately owned and susceptible to widespread, and frequent, clear-cutting (see Solutions 20).

Because of the history of national forest lands in the Blue Ridge, their trees are often eerily similar in age, like a nation composed almost entirely of adolescents. And the history raises a question: since few, if any, untouched forests remain, aren't "natural" forests here a paradox, like a natural rose garden or a natural wheatfield? The "What does 'natural' really mean?" question poses some complex choices for the future.

One far-reaching decision now confronting national forest managers, for example, is whether to try to regenerate declining oak forests. They provide a rich crop of acorns to support wildlife, but perhaps natural succession should replace the oaks with other tree species instead. Given the disruptions already caused by the introduced chestnut blight, which approach is more "natural," and which will promote biodiversity in the long run?

The happiest solution, and the luckiest, would be to find a way to begin restoring the chestnut itself to the Blue Ridge. (Forest Service funding for chestnut restoration research is, however, negligible: less than 0.25 percent of its $188 million annual research budget, and far less than the $91 million portion of the agency's research that directly benefits the wood products industry.)

But if restoring national forest and national park lands to pristine or "untouched" condition is an elusive goal, this related idea is easier to pursue: the forests could be allowed to return, instead, to a naturally evolving condition, in which ecosystem forces predominate, as free from outside influences as reasonably possible.

This view, suggested by some but not all scientists, concedes that human management—controlled burning, limiting deer herds, the removal of exotics, and the reintroduction of native species—will often be necessary. It's philosophically tempting, perhaps, to just "let nature take its course." But for now, given the destabilizing disturbances of the forests' immediate past, that course can lead toward ruin. Restoration will take time and help.

Old growth—the 0.5 percent to 3 percent of it that may still exist—is nearly all on public land. Imagine that the Blue Ridge national forests and the national parks were managed in the future so as to return them to a naturally evolving condition. They would still total "potential" or "future" protected old growth of less than a third of the Blue Ridge forest ecosystem. In that modest remnant, the interplay of disturbance, decline, and regrowth could lead to an increasingly natural mix of ages and kinds of trees.

At least some, perhaps most, of the other plant and animal species that inhabit a relatively undisturbed ecosystem would also reemerge over time. With prescribed burning where it is needed—and ignoring for the moment the potential effects of air pollution, exotic species, and climate change—these forests could resume something like their naturally evolving life patterns over the next 150 to 400 years.

The Blue Ridge embraces all or part of seven national forests—2.6 million acres for which Americans hold the deeds of trust (plate 7). They are storehouses of wood pulp and saw timber whose market value will climb during the next few years. They are the first step in restoring a fraction of the mountain forest ecosystems eliminated during the past century. And they are a wild, green "pleasuring ground"—Ulysses S. Grant's phrase—for a fast-growing population of recreationists, nature seekers, and tourists. The future of the national forests is, in many ways, up for grabs.

As logging and farming cleared the mountains early in the twentieth century, what we now call the "ecosystem services" that the forests conferred went with them, especially the land's ability to retain topsoil and moisture. Trees anchor soil, and soil holds water. That slows the runoff from mountain rainstorms, preventing floods, and maintains the flow of streams during dry periods.

Government reports documented accelerating erosion that choked navigable rivers with mud, generated vast floods, prolonged the effects of dry periods, fouled local water supplies, and threatened a future shortage of timber.

All this helped lead to the inauguration, in 1911, of the national forest system, and Congress authorized funds for the U.S. Forest Service to begin buying property. The first national forest, the Pisgah, was established near Asheville in 1916.

The most striking change across the Blue Ridge since that time has been its afforestation, the regeneration of trees on both public and private land. About 84 percent of the Blue Ridge ecosystem now has tree cover, which in turn has supported a resurgence of some kinds of wildlife, including turkey, deer, and bear populations.

The Forest Service, which manages the Washington, Jefferson, Pisgah, Nantahala, Cherokee, Chattahoochee, and Sumter National Forests on the Blue Ridge, is a branch of the Department of Agriculture. Its job description today is complex and full of potential contradictions. Here are some of its planning guidelines under various federal laws:

- Establish the amount of timber that can be cut and sold without jeopardizing future timber production.
- Recognize that the national forests are ecosystems and their management for goods and services requires an awareness and consideration of the interrelationships among plants, animals, soil, water, air, and other environmental elements within such ecosystems.
- Provide for the safe use and enjoyment of the forest resources by the public.

- Encourage early and frequent public participation in planning.
- Respond to changing conditions of the land and other resources and to changing social and economic demands of the American people.

In practical terms, the Forest Service must interpret such guidelines so as to attempt to respond to—or find a way to deflect—pressure from Congress, local citizens, environmentalists, the timber industry, off-road vehicle enthusiasts, skiers, grouse hunters, deer hunters, fishers, hikers, bikers, miners, rock climbers, and hang gliders, among others.

Federal law requires a ten- to fifteen-year "Forest Plan" for each national forest to try to reconcile all these interests, but the plans are open to major changes through appeal, revision, and amendment. New forest plans for the Blue Ridge—spelling out the future management of the Sumter, Cherokee, Chattahoochee, and Jefferson National Forests—were scheduled for the late 1990s. The George Washington, the Pisgah, and the Nantahala forest plans were finished a few years before.

It is these plans, and challenges to them, that set priorities on such matters as the kind and quantity of logging, the preservation of old-growth forests, recreational uses, and the health and stability of the ecosystem and its inhabitants.

Until very recently, growing trees and selling them to industry has been a controlling preoccupation of national forest administrators, under a concept called "sustained yield—multiple use." The policy has resulted in an economic boost for the wood products industry, its employees, and some local economies: a rich harvest of Blue Ridge timber over the years, often sold at less than the cost to taxpayers of administering the program. It has also resulted in hundreds of thousands of acres of clear-cuts and pine plantations and thousands of miles of forest roads.

It's beyond question, though, that national forestlands are now in vastly better condition than they were before the Forest Service acquired them. Heavy logging may have been controversial since its inception, but edge effects, biodiversity, and sustainable ecosystems were not much discussed, especially in the Southeast, until recently. Most citizens seemed to approve, tolerate, or be willing to ignore roading, logging, and mining in national forests. They were, in the service's sunny slogan, the "land of many uses."

During the 1980s, the pace of logging in its Blue Ridge forests jumped sharply. The Forest Service proposed decades of increased logging, and replacement of natural regrowth with "pine farm" plantations for future commercial harvests (fig. 18). Then the pace slowed, at least temporarily. This turn in policy was the result of public pressure, new worries about

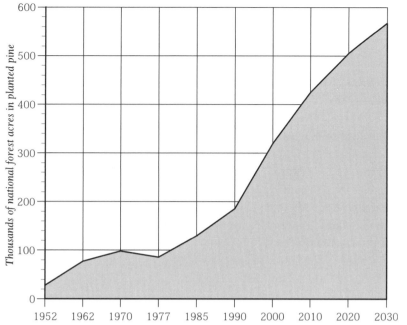

Figure 18. *What the Forest Service had in mind for the next century.* At the end of the 1980s, the Forest Service recommended these steady increases in acreage to be logged and replaced with plantations of fast-growing, marketable pine on national forests in Georgia, Tennessee, North and South Carolina, and Virginia (the figures also include national forests in these states that are not in the Blue Ridge). In the mid-1990s, the agency scaled back these plans for logging and pine plantations, at least temporarily. (*Source*: Department of Agriculture, *Fourth Forest*, 347–69)

the health of forest ecosystems, and shifts in our uncertain political breezes in the mid-1990s.

With a change in national administrations at that time, a new overarching policy called "ecosystem management" was announced. The implications are far-reaching. Priorities are supposed to move toward sustainable ecosystems, though logging and the rest of the catalog of "multiple uses" are still part of the forests' future under the new policy.

Many Forest Service employees have welcomed, and worked hard to implement, this change in outlook. But an agency involved so heavily in industrial forestry does not morph into an ecological health care system overnight. Forest Service personnel now wait with varying degrees of enthusiasm to see how the new policy will be honored during coming years and whether it will survive the political counterpressures it has provoked.

Critics among scientists sense that the new attachment to ecosystem management is tentative at best and mere lip service at worst.

Others believe it is supple. Plant pathologist Martin McKenzie, for example:

"Ecosystem management will become the way of handling forest issues," he says. "Even if there is a change in government, I don't think there will be a move back, at least in the East, to 'total product.' . . . I'm sure you will see a change to biodiversity issues, to valuing species for wildlife, for recreation, for environmental reasons."

In any case, of the acreage identified as "possible old growth" in national forests in the Southern Appalachians, more than a third was still designated as open to possible future logging as recently as 1995. The most recent Forest Service guidelines explicitly provide that remaining and "future" old growth—forests qualifying as "old" as the years pass—may still be classified as "suitable for harvest." As if to demonstrate that commitment, the agency has fought hard to continue logging in old-growth forests in some places around the country.

The same government that bestows "ecosystem management" on the national forests quickly changes course, depending on its perception of public opinion and other forms of political leverage. Despite its new policies and outlook, the Forest Service was forced by Congress in 1995 to carry out "salvage" clear-cuts that included old-growth forests in the West. President Bill Clinton at first vetoed, then signed, the measure. Then his administration shifted again, according to former Forest Service chief Jack Ward Thomas, trying to halt timber sales that were already contracted for.

Clear-cutting has been one of the Forest Service's most controversial practices. The agency adopted the widespread use of "even-aged management" in 1963. This term includes clear-cutting as well as kindred techniques. They are all a departure from using selective or group cuts, in which single trees, or a small number, are removed from an area.

Since then, about 15 percent of national forest land in the Southern Appalachians has been cut over using "even-aged management." At the same time, many cut-over stands—about 250,000 acres in the Blue Ridge between 1980 and 1996—became plantations of single species, usually fast-growing pines.

In debates over clear-cuts in particular, foresters in government, industry, and academia divide roughly into two camps. Many think in terms of a sustainable yield of timber, which they refer to as a "renewable resource" because trees of some kind usually grow back after logging. Others are concerned about sustaining the whole forest ecosystem, not just the timber supply. On issues such as clear-cutting, the two camps often talk past each other.

"Clear-cuts look bad, but they are not an environmental disaster at all," says John Seiler, a Virginia Tech forest biologist. "Nobody gets upset when the corn gets harvested in the fall. It looks just as bad." In the eyes of industrial foresters, objections to clear-cutting seem like a kind of incomprehensible, wimpy sentimentality that ignores the economic value of the timber. When older trees are cut in the Blue Ridge, they point out, a green explosion of new growth and a surge in some species of wildlife populations result.

But clear-cuts are antithetical to ecosystem health, other biologists argue. Stephen Trombulak, for example, writes that abrupt reductions of biological diversity can persist for centuries after catastrophic disturbances, including clear-cuts. The new growth that follows them is impressive but less diverse.

Despite decades of controversy over clear-cutting, research about its aftereffects on the whole ecosystem—not just supplies of timber for logging markets and wild game for hunting—is relatively new. One approach has been to examine how clear-cuts can affect abundant, rather than rare, kinds of plant and animal life.

Salamanders, certainly, are abundant. Seldom glimpsed by most of us, they may be the commonest form of vertebrate, or "backboned," life in the Blue Ridge. The mountains support a great diversity of salamanders, from the 2-inch-long pygmy to the 2-foot-long Giant Hellbender—thirty-five different species in North Carolina alone.

If you collected all of the salamanders from a hundred acres of forest in a big basket and weighed it, they would probably account for more "biomass"—that is, more living animal material—than all the other vertebrate predators, from birds to bears to chipmunks, combined.

Salamanders are important to the balance of natural systems. They dine on great quantities of insects, for one thing, and they in turn are important food for other predators. Because of their diversity and abundance, they are also a significant indicator of the health of the whole ecosystem.

Research in and near the Craggy Mountains north of Asheville compared salamander populations in newer clear-cuts with those in clear-cuts fifty years old or more. Biologists found that "clear-cutting strongly depletes local populations of salamanders and reduces local species richness. We estimate that about 75–80 percent of salamanders in mature stands are lost following timber harvesting by clear-cutting." About fifty to seventy years are probably required for populations to fully return.

Clear-cutting in national forests in western North Carolina alone resulted in the loss of nearly fourteen million salamanders each year, they calculate. The cumulative effect of clear-cuts over time has been that regional populations are a quarter of a billion salamanders lower than if the forests had not been clear-cut.

This study was not the last word on the subject. It attracted critical comment, as scientific work often does. "The fate of salamanders on clear-cuts is something we need to continue to investigate, but it is important to stress that their fate is unclear at present. We know they disappear from clear-cuts, but that is all we know," one set of critics wrote.

Other research has concentrated on the effect of clear-cutting on wildflowers and other understory plants. It paired old-growth sites from North Carolina to northern Georgia with immediately adjacent sites that had been clear-cut forty-five to eighty-seven years before.

Even after this much time had passed, the diversity of the plants was severely reduced on the clear-cuts. On average, clear-cuts had only half as many species, and they covered only a third as much of the ground. There was no recovery with age: "If anything, both richness and cover appeared to be decreasing."

This work, too, drew criticism. The title of one rejoinder was "The Effects of Clearcutting on Herbaceous Understories Are Still Not Fully Known." That uncertainty is at least one conclusion all seem to share, for salamander populations, plant diversity, and the recovery of the rest of the forest after it has been clear-cut. It forms a powerful—but not, so far, decisive—argument against continued clear-cutting and other kinds of even-aged management.

Citing "public and scientific concern," the Forest Service now predicts that clear-cutting—but not the other kinds of "even-aged management"—will decline on all national forests in the United States by 81 percent during the next half century. But the total yearly volume of timber to be removed will only fall by about 25 percent by 2045, the agency estimates.

Scientists in and outside the Forest Service have suggested experiments to test this idea: selective logging of individual trees or small groups of trees, as opposed to clear-cutting, may at times mimic natural disturbances and be compatible with natural systems. Others argue that roads and other disturbances would in some ways be even more intense were this kind of "selective cutting" to take place. Long-term research is essential to testing such ideas. The question that logically precedes the research is which set of concerns—biodiversity or logging—is paramount? Which is the tail and which the dog?

Logging requires roads, and roads have taken an especially heavy toll on the ecosystem. Nearly all of the Blue Ridge, including the national forests, is overlain by a dense spiderweb of old logging roads, old logging railroad grades, and Forest Service roads—so many that they have never been thoroughly mapped or cataloged. About 6,000 miles of maintained roads under Forest Service jurisdiction now blanket the Blue Ridge. They open the forest to edge effects, invasions of exotic plants, erosion, and silt-polluted streams.

Erosion from logging roads can contribute to the ruin of watersheds, a fact documented by Forest Service research that began in the 1930s. That research helped lead to the development of "best management practices" meant to guide road engineering on national forestlands. But even the most carefully laid out and maintained roads mobilize soil that can end up in streams. The cumulative effect of roads, both good and bad, has been massive in watersheds like that of the Chattooga River, one of the two federally designated "wild and scenic rivers" in the Blue Ridge. The 57-mile-long Chattooga originates in North Carolina and forms part of the border between South Carolina and Georgia. Two-thirds of its watershed lies within the Chattahoochee and Nantahala National Forests.

Clemson University professor D. H. Van Lear flew aerial surveys over the river during a recent year-long study of erosion and its effects. He often saw underwater plumes of sand, plainly visible following storms, moving from tributaries into the main channel and along the river bottom, obliterating fish habitat.

"I'm certain there would be many more trout if there was less sand, fine sand and silt in the river," he says. Van Lear and his colleagues found that 80 percent of this kind of pollution in the Chattooga comes from dirt roads on private, county, and national forestlands—of which the last were probably in the best shape. An additional 9 percent of the pollution is generated by timber harvest access roads and loggers' skid trails.

"As one of the few remaining free-flowing rivers in the Southeast, the Chattooga is famous for its boating, fishing, wildlife, scenery and aesthetic value," the study notes. It concludes that "it may take decades or even centuries for the river bottom to approach full recovery," assuming that the erosion is slowed.

As noted in an earlier chapter, roadless areas account for only 3 percent of the land area of the entire Southern Appalachian region, including national parks and forests and state parks (fig. 17). Federal criteria define "roadless" as having less than one-half mile of improved road per 1,000 acres, and some logging roads are not included in the calculation. More than a third of the total of the "roadless" area of the entire Southern Appalachians is in a single place: Great Smoky Mountains National Park.

Forest Service planning over the years has been geared toward carving new roads, even in the few remaining roadless areas despite their extreme

SOLUTIONS 17

Old Growth and Forest Policy

[Robert Zahner is professor emeritus of forestry at Clemson University and a self-described "forest activist." His family has lived on Billy Cabin Ridge, in the Cowee Mountains near Highlands, North Carolina, for three generations.]

Early on, I was a research scientist for the Forest Service. My job was to help increase timber production on the national forests. The service was moving very strongly into the industrial mode all over the South. When I began teaching forestry, we turned out a lot of graduate student projects on the same subject: increasing production. But my interests began to shift toward the relationship between forest management and biological diversity. When I arrived at Clemson, the Forest Service was placing tremendous emphasis on clear-cutting hardwoods and converting hardwood stands into pine, and many of them were even in the highest mountain areas.

How has Forest Service policy on the Blue Ridge changed recently?
Local issues and local people have succeeded to some degree in getting the Pisgah and Nantahala National Forests to move away from clear-cutting, reducing the amount of land set aside for logging, and managing more of the forest for wildlife, biological diversity, and recreation. Their management plan is now something of a model for the other national forests in the Southern Appalachians.

Has clear-cutting been abolished, then?
Very little of it is done, but it is still available as a tool. Much lighter and more selective kinds of cutting, leaving several different ages of trees, and natural reseeding—no pine plantations—is a lot better for the forest and for diversity.

Is the Forest Service commitment to ecosystem management strong?
The progress is in administrative policy. It sounds good to the public, and the Forest Service gets a lot of PR out of it, but when you bring it down to ground level, where policies are carried out, there are many abuses, and it is not always well managed. Some even-aged logging sites, like the Big Creek timber sale near Highlands (North Carolina), are a big eroded mess. Too many trees were cut, and most of the trees left standing are so badly damaged they're going to die anyway.

What should happen now?
Any stands that can be classified as old growth should be protected from logging. In younger stands, we should only use selective cuts of larger trees and stop cutting small trees for pulp.

How much old growth is enough?
If it were realistic politically, I'd favor setting aside all national forests in the Blue Ridge for old growth, simply because public land—the only place where old growth can be protected by law—is such a small percentage of the landscape. Timber can, of course, be harvested on private lands.

Don't we need some younger forests for bear, deer, and grouse?
No, we don't need it for the animals. They do very, very well in old growth, and they always have. When you cut trees for temporary population increases in those species, you do it for the hunters, so more of the animals can be killed. It isn't for the animals.

What about proposals to reserve all public forest lands in the Blue Ridge for recreation, research, and the protection of biological diversity?
Eventually, it may have to come to that, but that's a really long-range goal. Even those proposals envision some moderate logging in buffer zones on the periphery that have already been heavily cut over in the past—but not in core biological areas.

What about roads?
Definitely no new roads, except for emergencies. We don't need any more logging roads. All the forest that could conceivably be logged is already roaded and over-roaded. We need to start closing large numbers of roads. That will help biologically, and it will cut down on poaching.

How stable is "ecosystem management" policy on the national forests?
Only as stable as Congress and the timber industry will allow it to be. There's definitely a backlash against enlightened forest policy now, even in professional forestry groups, where I had hoped that thinking was evolving beyond looking at forests as commodities. Timber industry scientists and forest service scientists who are under political pressure are advising Congress to do away with ecosystem management—just as we are finally beginning to look at the forest as a whole and not just as timber.

Where's the support for that view, then?
Urban people and their representatives in Congress will have to defend ecosystem management in the forests for recreational uses, for its inherent values such as biodiversity, and for water quality in rivers and reservoirs.

Source: Zahner interview.

SOLUTIONS 18

Sawmills and Clear-cuts

[Eldon Bradley owns and runs a sawmill just across Route 60 from Buffalo Creek that his grandfather, James Pamplin Bradley, established in 1908, on the margin of what would later become part of Virginia's George Washington National Forest. Outside his small log-cabin office recently, someone was welding a broken knuckle-boom loader, and a band saw whined through piles of poplar and red, white, and black oak bought from independent loggers. It would become furniture, molding, and trim.

"You have a good year," he said, "then you'll have a year and a half where you're doing darn well if you can just pay the bills. Right now, the whole industry is overstocked with timber and prices are dropping fast."]

How has the business changed since you were growing up here?
It was bigger timber then. There was a lot of original growth timber, and there were logs from the national forest. I haven't bought any timber off the national forest in four or five years. There's very little coming up for sale in this area.

Do you care whether they cut timber on the national forest or not?
Timber to me is a crop. I certainly don't advocate clear-cutting little healthy poplars and oaks, which was done at one period on the George Washington. They went through that phase, and it was awful. They shouldn't have been doing it in the first place. But if timber is mature, I hate to see it wasted.

What's wrong with clear-cutting?
I have seen very few places in these mountains that ought to be clear-cut. Very few. If you get back up on top of these ridges where it's rocky and the land is poor, if you clear-cut it, you open it up to erosion.

rarity. As recently as 1990, the agency planned to shrink large roadless areas on southern national forests by 24,000 acres during the ensuing half century and to expand its inventory of roads by 1,500 miles.

The Cherokee National Forest alone built or reconstructed 292 miles of road from 1988 to 1995, mostly to access timber sales. The forest plan for the George Washington National Forest specified that on more than 10 percent of its still unroaded sections, "projects may be scheduled that might substantially change the roadless nature of these areas." The plan projected 50 to 80 miles of new road on the forest between 1993 and 2003 and stated that "additional roads may be needed."

The most recent national long-range plan, never issued in final form, stated that the Forest Service's "future position" is that some roadless areas would still be available for timber sales and thus conversion to "roaded"

What about big saw timber?
I think it ought to be selectively cut. But around here you don't really find that type of timber. A gentleman I know near here would bring in a load of logs a day, from the early fifties to the mid-nineties, selective cutting. That, in my opinion, is the way to handle 95 to 97 percent of the timber in this area.

But clear-cuts can move more timber out using fewer people in a shorter time. They are said to be better, economically.
Better for who? If you cut one of these slopes up here, the first thing that's going to come is the blamed briers, thorns, locusts—the junk stuff—then you'll see poplar and some oak—you're talking a hundred years before you even see merchantable timber. In some places, it never will come back.

What's going to happen to the sawmill business here during the next few years?
It's going to get tougher all the time. I have seen the size of the timber get smaller and lower quality every year.

Are we going to have to continue logging on the national forests?
Absolutely. Any time you take part of your raw material off the market, you're going to have to go somewhere else. You have to go to private landowners. Somewhere down the road it's going to have to have an effect on prices. So I'm in favor of cutting, on a limited basis.

Source: Bradley interview.

status. But in 1998, the agency proposed a controversial eighteen-month moratorium on road building in roadless areas. It also called for public comment on a new national policy for slowing its road-building program and closing and revegetating some existing roads. The rationale is that the road system not only causes ecosystem damage but is too expensive for the agency to maintain.

When I was young, my stepfather, Don Hassler, made his living as a logger in the national forests of the West, in the Siskiyou Mountains and along the Eel River. The scene was chain saws and mill ponds, trailer trucks, trailer parks, and long weeks of idleness and tension when the work ran out. The threat of unemployment applies a powerful torque to politics in such communities. And on the Blue Ridge today, the decisive argument in favor of clear-cutting or any logging in national forests has to do with economics, not ecosystems: "We need the jobs," the argument runs.

There is no simple way to juxtapose those jobs against the value that intact natural systems have for humans or for their own sake. One thing economists can do, though, is to estimate just how essential logging in national forests—as opposed to private lands—may be in creating future jobs and supplies of wood. Their calculations are worth considering carefully.

The employment value of logging goes well beyond the paycheck of the person with the chain saw. Many more jobs are created as the trees are processed. About 40 percent of the wood from national forests in the Blue Ridge goes to pulp mills to become cardboard, writing tablets, paper cups, milk cartons, and other products. The bulk of the rest is saw timber, mostly hardwood such as oak and poplar, for wood products.

How important are national forest trees for creating jobs? Figures for the Blue Ridge alone are not available, but in the Southern Appalachian region as a whole:

- Only 10 to 12 percent of the commercial timber harvest came from the national forests during the 1980s and early 1990s, and the percentage has been dropping.
- About one-half of 1 percent of the region's total payroll was generated by logging national forest trees during that period. The figure includes pulp and paper, furniture, and all other wood-related employment.
- Adding up the direct effects, and indirect "ripple" effects, generated by commercial logging in the Southern Appalachian national forests accounts for a roughly estimated 1.2 percent of total employment in the region, and 1.1 percent of total economic activity, according to Forest Service economist David Wear.

These jobs would not disappear, even if all commercial logging stopped in Southern Appalachian national forests tomorrow. Instead, the profit motive would beckon suppliers to step in to meet the demand from private lands, which already provide nine-tenths of the timber in the region.

The market for a few types of wood such as red oak would be "strongly affected" by the withdrawal of national forest timber. But overall, wood supplies would be stable: "Actions taken by the [national] forests may not influence prices at a market level because there are adequate substitute sources of material in the region," Forest Service economists say.

Factors far removed from national forest policy such as automation, the increased use of synthetic materials, and alternative supplies of timber may diminish future southern employment in wood products industries. The

Forest Service has projected that from 11 to 40 percent of those jobs will disappear between 1990 and 2030 (fig. 19).

So concerns over jobs tied to logging in national forests should not be disproportionate to the situation: a possible relocation—not a "downsizing"—of perhaps 0.5 percent of the payroll and perhaps 1.2 percent of all employment, in the Southern Appalachians. And this in an industry where jobs are likely to decline substantially over the coming decades anyway.

Those figures, we should remember, are for the whole region. By contrast, local economic dislocations would be likely if logging ceased in national forests altogether. Logging activity is not uniform throughout the Blue Ridge. In a group of six counties from Asheville to the Georgia border, for example, national forests accounted for half of local timber production during the 1985–94 period. If they were closed to logging, and if the timber on private lands proved to be far enough away, mills might shift some jobs to other locations or move their operations nearer the source of supply, the economists say.

If local communities are to be weaned from dependence on national forest timber, an extended timetable will make the changes easier to adjust to. Zigzag policy making, on the other hand, breaks faith with those communities and provokes resentment. It turns the U.S. Forest Service, and the federal government generally, into a fickle economic partner.

Meanwhile, another activity in the national forests and in the Blue Ridge economy—recreation—booms. Growing populations, and their growing appetite for outdoor recreation, may soon crowd national forests that once seemed big enough to accommodate every use.

According to the federal/state, multiagency Southern Appalachian Assessment:

- "Large urban areas are expanding to the edges of public land in the Blue Ridge. . . . One result is a high density of use at the outer edges of public forests and parks. . . . As the population centers grow, high density use patterns will creep toward the center of the mountain ranges."
- "Public land provides key resource attributes for future generations while providing enjoyment now. . . . Nearly one-fourth of the area in national forests is in landscapes categorized as 'outstanding' or 'distinctive.'"
- "Nature and the outdoors are being aggressively and successfully marketed. In addition, telecommunication from remote sites allows

SOLUTIONS 19

View from the Top: Scramble Ahead

[Jack Ward Thomas was appointed chief of the Forest Service by the Clinton administration in 1993—the first such appointee trained as a wildlife biologist. He stepped down just before the national election of 1996, saying that the agency was being whipsawed by politics. He now teaches at the University of Montana.]

How did the Forest Service get involved with "ecosystem management?"
The final recognition that we were going to need it came out of the Pacific Northwest, where we dealt with the spotted owl issue, which was not a spotted owl issue at all but an old-growth issue—a significant portion of an ecosystem that was shrinking and being fragmented by timber sales.

Suddenly as we began to have to deal with one threatened or endangered species after another, it became clear that we would have to take a broader view over a longer period of time. Now, the de facto policy has become—though nobody's quite said so openly—that biodiversity preservation is the overriding objective of federal land management.

People moving into rural areas near the national forests now have considerably more concerns about aesthetics and wildlife and other issues—I think more of a "not in my backyard" reaction than anything else. They began to tell us they were more interested in other things than maximum timber production.

Do you see a fairly stable future for management of the national forests?
We're one law, one lawsuit, one budget, one insect outbreak, one introduction of some exotic pest, one loss of a pesticide, one election, one change in public opinion away from instability all the time. The old dream of the regulated forest is simply something that we cannot fulfill. The only time stability was ever possible was when we were cutting against the virgin forest.

What will the role of the timber industry be, then?
The timber industry and the people that depend on it see an opportunity in our "paralysis by analysis" routine that has made it very expensive to do something as simple as selling national forest timber. They will push for granting some powers and rights over those lands back to the states or to private industry. We'll have to see what happens. Those ideas have come and gone a number of times, and we're at it again.

What would allow the Forest Service to do the best job of ecosystem management?
I'd give administrators more authority over how to allocate funds. Right now, the ecosystem has no constituency, except for some scientists and environmentalists. The fish and wildlife folks are interested in hunting and fishing, the timber folks are interested in timber, the recreation people in recreation, and

so on. So when the Forest Service manager is sitting there trying to balance budget and staff, and it's all specified in a number of little line-items—particularly when it has earmarks of various kinds by Congressmen seeking to give out special favors—it's very difficult to put together any kind of coordinated program. Conservation is for the long term. . . . There is too much politicization of the Forest Service and too much erratic intervention by politicians and their appointees.

It's been proposed that the whole Blue Ridge be turned into the equivalent of a wilderness area or a national park, for ecosystem management, nonharmful recreation, and scientific study.

It would fit all right with ecosystem management at a very large regional scale, but I don't know that there is much public or political support for it.

Does the idea appeal to you?

Not particularly. If, for example, you're interested in hunting turkey, deer, and grouse, that's not the best kind of habitat management. National forests aren't the only areas available for hunting, but they make up most of it. But if biodiversity is your priority, perhaps that would be a good idea, especially for species that tend to be associated with broad expanses of older forests. And there are certainly plenty of younger forests being produced all the time in the surrounding region, on private lands. Ecosystem management implies that we need to consider larger areas, like the southeast region, rather than just one ranger district or one national forest at a time, for planning.

What would you like to see happen?

If our public policy is to protect as many endangered species as we can, and if the priority really is biodiversity protection, things will move increasingly toward preservation. It isn't a stated nor well-thought-out policy but a series of piecemeal responses to events that have been strung together, without much thought, that have put us on the "yellow brick road."

Is that a good road to be on?

Depends on what society's objectives are. That gets to be a matter of personal choice. My mind-set runs thoroughly toward multiple use for those lands. I'm not so sure that drawing a line around the national forests and walking off from them—particularly since the forests in the Southeast are not pristine "natural" forests—is good. It will have serious effects on local cultures and communities that have used the forests for centuries.

We're cutting timber in the Southeast faster than we are growing it now, on private land. All of a sudden, sometime in the future, someone's going to start looking back at the national forests to take up the slack while private lands catch up again.

SOLUTIONS 19 (cont.)

What role will recreation play in the future?
It's constantly increasing on the national forests. We're experimenting with user fees, because the trick to it is that recreational use is something we're just not budgeted to handle. People have said they'll be willing to pay substantial amounts, but they have also said they want the money to support those recreational uses on the forests. They don't want the money to be handed back over to the national treasury to be reallocated by politicians. That's a revolutionary concept, though, because traditionally all government revenue goes back to the treasury.

How can the Forest Service deliver a working ecosystem to the next couple of generations, in the face of challenges from a growing population, exotic pests, and air pollution?
We're going to be in a constant race to control those problems with new technology. At the same time, we've got a Congress that doesn't even seem sure it wants people to know how to read and write, judging by the way they've cut research funds. If you think research on sustainable resource management is expensive, consider the long-term cost of ignorance.

It's going to be a struggle to maintain the research to combat these problems into the future. This is going to be a scramble at best, requiring continuous, rolling, adaptive management. The system we have now is simply not conducive to that: it takes us a year to decide to go the bathroom.

Source: Thomas interview.

people greater freedom to work away from home, even in natural-appearing settings. This trend will likely continue, putting increasing pressure on nature-based recreation settings and facilities."

The Census Bureau projects that by 2025, the populations of Virginia, Tennessee, South Carolina, North Carolina, and Georgia combined will increase by about a third—an additional nine million—over the 1995 figure.

Forest Service figures show a continuing and steep upward trend in the number of recreational visitors to Blue Ridge national forests, from 5.8 million in 1970 to 13.4 million in 1996—an increase of nearly 300,000 each year, on average, during that period.

"Activity opportunities are abundant," the Southern Appalachian Assessment states, but roads along rivers that provide access to fishing and camping, interconnected trail networks for biking, horseback riding, off-road vehicle driving and hiking, and trails and roads to such places as waterfalls and scenic overlooks are "becoming crowded" (figs. 20, 21, 22).

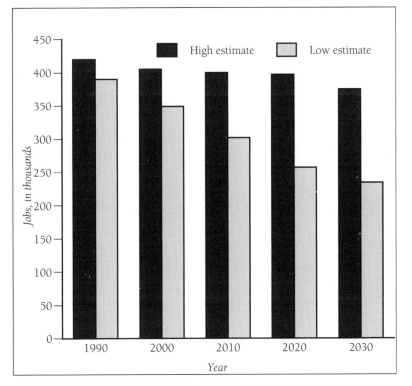

Figure 19. *Wood-related employment in the South, 1990–2030.* Forest Service projections for all wood products, logging, and pulp and paper industries indicated that the jobs they provide would decline by 12 to 40 percent between 1990 and 2030. The estimates varied with potential changes in imports, exports, technology, logging rates, and other factors, but all scenarios predicted job declines. (*Source*: Department of Agriculture, *Fourth Forest*, 461)

In the South, the percentage of the population that participates in out-door recreation is rising steeply, even while the population itself expands. Total recreational visits to Shenandoah and Great Smoky Mountains National Parks did not increase much during recent years. But tent camping at the parks rose 11 percent just from 1989 to 1995, and backcountry camping was up by 27 percent during the same period.

In 1972, only 3.6 percent of southerners who responded to a survey said they participated in primitive camping; in 1992, the percentage had more than tripled, to 12.9. Participation in developed camping jumped from 8 percent to 18.6 percent; the number of day hikers also increased sharply, from 2.8 percent to 19.7 percent. Even the wilderness areas are beginning to show wear and tear, Forest Service managers say, and new polices limiting the size of groups have been imposed in some areas. The number of

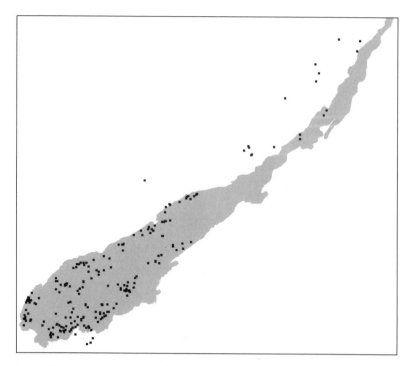

Figure 20. *Recreation sites in the Blue Ridge where capacity is exceeded during peak week-ends. (Sources: Keys et al., Ecological Units of the Eastern United States, CD-ROM; Hermann, Southern Appalachian Assessment GIS Data Base, CD-ROM)*

recreation spots in the Blue Ridge marked "capacity sometimes exceeded" in government tallies is increasing.

U.S. Forest Service stewardship on its big patches of the Blue Ridge has been long, and its reams of plans and procedures bespeak order and continuity on into the new century. The reality is somewhat different, however (see Solutions 17, 19).

A walk up Mount Pleasant, in Virginia, offers perspective. The trail and the mountain are part of a recently declared national scenic area. Many slopes here are thick with spindly, brushy young forest that is gradually reclaiming a logged-off and eroded landscape privately owned until the 1970s. The route also winds through groves of big old poplar, oak, and pine.

On top, the wide view from Mount Pleasant's granite cap forms a tableau of Forest Service management during the past century. It also inspires guesses about what may occur in the coming one, as the agency is pushed toward, or away from, concern for a sustainable ecosystem.

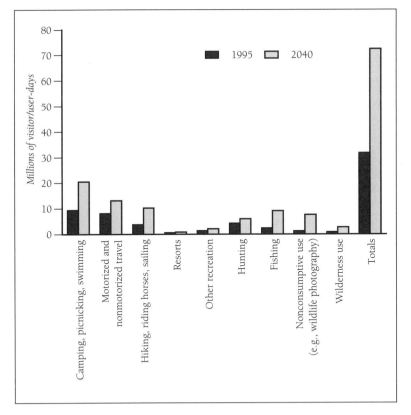

Figure 21. *Future recreational demand on southern national forests.* The Forest Service projects more than double the total number of 1995 "visitor-days" on southern national forests by the year 2040. (*Source*: Department of Agriculture, Forest Service, *Program for Forest and Rangeland Resources*, 1990, table E-8)

Flanked on both sides by the Blue Ridge and its outliers, a broad valley stretches southwest. Much of it lies within the boundaries of the Pedlar Ranger District, one of six that make up the George Washington National Forest.

To the northwest, a uniform gray-green patch of white pines contrasts with the hardwoods below the saddle of Cold Mountain. That is stand 52 of compartment 1191. The hybrid trees were planted in 1983 to reforest what may have been an old pasture.

"There was a time when the philosophy was to grow as much biomass out there as you could, and pine will grow more than oak," Forest Service silviculturist Sharon Mohney says. "Currently, we recognize that oak is a highly valuable part of the ecosystem for wildlife. Now when we harvest, we usually want oak to come back."

Panther Mountain defines the east side of the valley. Clear-cuts from a

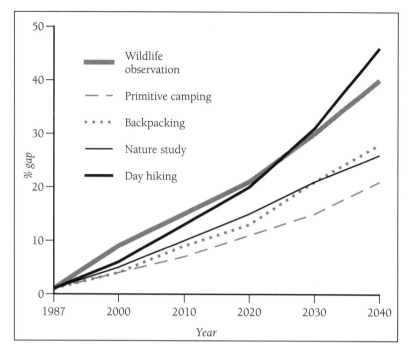

Figure 22. *The coming gap between recreational supply and maximum demand in the South.* Forest Service projections indicate that demand for nature-based recreational trips will increase in the future but the supply of recreational opportunities will increase much more slowly. The desire for camping, wildlife observation, backpacking, nature study, and day hiking trips will sometimes be unfulfilled because of higher prices and increasing congestion at recreation sites. This creates the projected "gap" between supply and demand. (*Source*: English et al., "Outdoor Recreation and Wilderness," 33–36)

timber sale there in 1989 are visible where 85 acres of white pine, yellow poplar, and oak stood. They produced 116,900 cubic feet of pulpwood and 766,000 feet of saw timber. It is possible that, by now, those trees would have been more than a century old. Logging old growth is not allowed for now, though, without at least considering its impact on regional biodiversity.

Over a ridge to the southwest is a pocket of trees that is, Mohney says, "old growth by anyone's definition," along the Appalachian Trail at a place called Little Irish Creek. A decrepit Forest Service sign there explains the value of those ancient pines, oaks, and hemlocks in terms of how many houses could be built from them. "That was the manner of the times," Mohney says. "If we were doing that sign today, we'd probably be a little more politic in how that site was described."

Roughly half of the valley lands are privately owned. Of the national forest portion, nearly all is designated as "suitable for timber production."

SOLUTIONS 20

Private Land, Public Issues: Chip Mills

More than three-fourths of the Blue Ridge's timberlands are in private hands. They are at least as important to sustaining the ecosystem as the national parks and forests: 62 percent of occurrences of populations of endangered species in the Southern Appalachians, for example, are on private lands.

Unlike some other states, in the Blue Ridge private-land clear-cuts and their effects on soils, streams, and wildlife are largely unregulated, despite their effects on off-site soils, streams, and wildlife.

And on private lands, the pace of clear-cut logging is accelerating, often to feed the appetites of "chip mills" that reduce logs to squares about the size of snack crackers, to be processed into pulp. Chip mill capacity in the Southeast is increasing.

One of the few attempts so far to assess the impact of chip mills on the forests, and on the pace of logging in the mountains, occurred in 1993. At that time, permit applications for new barge facilities on federal land in Tennessee were under consideration, along with the impact of three new chip mills to be served by the new shipping.

Research by the federal agencies involved found that the three new mills would double the annual acreage clear-cut each year. This would in turn generate adverse effects on wildlife in general and endangered species in particular. Their report emphasized that the states do not enforce "best management practices" that might diminish the impact of logging. The permits were denied.

Dozens of chip mills operate in the Southeast. Many new ones are proposed or under construction throughout the region.

Sources: Southern Appalachian Man and the Biosphere, *Southern Appalachian Assessment,* 4:83, 5:36; Tennessee Valley Authority et al., "Final Statement." xxii, xxvi, 50–52.

Several timber sales were completed in the mid-1990s, and more are probable. The valley may also include a few stands of oaks and conifers identified as "possible old growth" on Forest Service databases.

The view stretches to the billowing stacks of the Georgia Pacific Corporation's Big Island pulp mill, nearly 20 miles distant. The mill turns out hundreds of tons of cardboard each day from recycled waste and wood chips—an unknown fraction comes from the national forests—and employs four hundred people.

A new forest plan for the George Washington National Forest, including the Mount Pleasant area, was completed in 1986. As a Forest Service document notes, the plan "did not gain public acceptance . . . and was the subject of 18 appeals." Those appeals, usually pitting ecosystem concerns against other forest uses, precipitated another complete revision that took seven years.

The last of the appeals to the plan was not resolved until 1996, which made the ten-year plan ten years overdue. In 2001, the process of pulling together the next plan for the George Washington will begin.

Whether the next century will bring more logging and logging roads or more wilderness, roadless areas, and protected old growth is an open question there and on the other national forests. "And politically," Mohney remarks, "it always will be."

The traces of something like a near miss are still there in the Snowbird Mountains after a hundred years, but they are mostly hidden now by brush and fallen trees. Near the 3,500-foot elevation on an unnamed ridge is the entrance to an old mine tunnel, blasted 75 feet back into the rock in 1887 by the Belding Lumber Company.

Local historian Marshall McClung has found old field diaries that explain the tunnelers' quest. They hoped to discover the extension of a rich finger of copper being mined and smelted 50 miles to the south. If they had found it, this part of the Blue Ridge might today look a lot like that one, which is known as the Copper Basin.

There, near the junction of Tennessee, North Carolina, and Georgia, a mining boom brought people, a measure of prosperity, and epic environmental destruction. Copper Basin ore was rich, and when railroads reached the area, large-scale mining began. As more and more mine pits were opened, the forests of the basin were cut over for fuelwood to roast the ore.

Especially in the years spanning the turn of the century, sulfur dioxide fumes and smoke poured out of the smelters, killing trees and other vegetation and poisoning the soil. A stagnant ocean of smoke, fumes, and fog, heavier than air, was often trapped in the basin. The mine holes, logging, and air pollution together yielded a 50-square-mile desert of barren, gullied land.

Federal agencies and a succession of mine owners planted millions of trees to try to stem the erosion and revegetate the poisoned soil, but most of the trees died. The devastation was still great enough even in the late 1980s to be clearly distinguishable on satellite images.

Since then, plantings have been more successful. Tree cover, the first step in the long, slow return of something like a healthy forest ecosystem, has been established. Now just a few of the sterile, red-orange scarscapes are left. They are only the most visible reminder of what has been called "one of the most severe and long-lasting occurrences of environmental degradation in the United States."

At the turn of the last century, as sulfur dioxide fumes from the Copper Basin smelters were killing forests and orchards—and, it was charged, making people sick, even in several counties over the Georgia border—Tennessee refused to move against the copper companies and disputed Georgia's right to interfere.

The legal struggle that followed was eventually heard by the U.S. Supreme Court, which ruled in Georgia's favor. It's worth noting that the majority opinion was a statement of values, not facts or law. It was delivered by the clearly indignant Chief Justice Oliver Wendell Holmes:

"It is a fair and reasonable demand on the part of a sovereign [state] that the air over its territory should not be polluted on a great scale by sulphurous acid gas, that the forests on its mountains . . . should not be further destroyed or threatened by the acts of persons beyond its control."

States still fight over intramural pollution in the Blue Ridge ecosystem, and Tennessee has played the role of outraged environmental advocate in one such conflict. The Pigeon River flows northwest into Tennessee, tinted brown from the tannins and lignins released by the Champion International paper mill in Canton, North Carolina, with North Carolina's official blessing. New technology was installed at the mill in the early 1990s, but it only removed about half the flow of pollutants.

The mill's owners threatened to close it down if more restrictions were imposed, and Tennessee threatened to sue North Carolina if they were not. Until an interim agreement was signed in 1998, Tennessee's official characterizations of this conflict displayed understandable ire.

Except that Tennessee strenuously defends its own need to export pollution into the mountains of North Carolina and Great Smoky Mountains National Park. During a recent fracas, Molly Ross of the U.S. Interior Department called Tennessee the most difficult state her office has to work with when it comes to resolving conflicts over industry, air pollution, and the national parks. "We have not run into as many severe procedural problems in any other state as we have in Tennessee," she has said. "Tennessee tops the list."

Recently, a new coal-fired industrial lime kiln complex was planned in Tennessee, just 35 miles from the boundary of the park. The Park Service's initial calculations were that the facility could dump as much as an additional 8 percent of nitrates on high-elevation soils and on streams already saturated with nitrates. Remote streams such as Noland Creek and Clingman's Creek in the national park, where you might expect water to be at its purest, already carry so much nitrate in spring that they approach the human health toxicity standard for drinking water.

The Park Service requested more refined tests to gauge the amount of pollution that might come from the kilns, but Tennessee officials refused to require them. Instead, they gave the plant the go-ahead. The Park Service strongly disputed this decision but eventually acquiesced, additional pollution or not. In return for that, the state of Tennessee agreed in writing to share information, promptly and thoroughly, about pollution and permits for new industries in the future.

But in less than a year, state officials reneged on the agreement, under pressure from the Tennessee Association of Business. "If the provisions of the current [agreement] remain in effect," the state's commissioner of economic and community development said, "it is my belief that prospective industries will tend to look at other states as more attractive locations."

Responding to the public outrage that followed, Tennessee governor Don Sundquist reinstated the old agreement with the Park Service temporarily and appointed a new special committee to "balance the needs of the environment and economic growth" in the park area. George T. Frampton Jr., a U.S. Interior Department official who was invited to participate, wrote back:

"Unfortunately, the state's approach to air pollution issues, both in recent days and over the years, has raised serious questions about Tennessee's commitment to protect national park resources. Tennessee's actions have suggested a failure to appreciate a basic truth which is emerging in communities across the nation: over the long term, economic well-being depends on environmental health."

Frampton's letter recalled, "In more than 30 permit reviews and related actions, Tennessee has rejected virtually every recommendation for measures that would assure that economic development occurs in a manner that also protects park resources."

The new lime kilns were fired up in the spring of 1997, at about the time that Tennessee officials vowed, once again, to cooperate with federal officials on sharing information about new sources of pollution in the future. The promise included a "bail-out clause," however. And fairly soon, the Tennessee governor was leading a hoped-for revolt against new federal air quality regulations. His administration, sounding a patriotic note, likened federal data on Tennessee's air pollution to the tea at the Boston Tea Party and invited fellow revolutionaries to "dump it in the harbor."

Many who lived in the Copper Basin found a kind of stark beauty in its vast, red-orange desolation, its 19-foot-deep gullies, gaping pits, and heaped-up tailings, calling the scene their "beloved scar." Exhibits in a small museum at the otherwise defunct Burra Burra copper mine in Ducktown also reveal this pride, and the history behind it. There, the faces of dozens of miners in the old photos remind us that while they were creating a man-made desert, these men also made a living.

Something like the same awakening came to the philosopher William James in a different part of the Blue Ridge, in a different era. Journeying among the primitive mountain farms of North Carolina in the late 1800s, he saw that "the forest had been destroyed; and what had 'improved' it out of existence was hideous, a sort of ulcer. . . . Ugly indeed seemed the life of the squatter."

Then he chanced to talk with the locals about their own sense of their surroundings: "When *they* looked on the hideous stumps, what they thought of was personal victory. The chips, the girdled trees and the vile split rails spoke of honest sweat, persistent toil and final reward . . . moral memories . . . of duty, struggle and success."

We are right to empathize when jobs are at risk because of measures taken to restore the ecosystem, or to prevent further damage. The thousand-plus workers at the Pigeon River mill can't all become whitewater rafting guides if they are out of work. There *are* new jobs to be had at the new lime kilns.

But there is a larger context to consider when the "jobs versus environment" issue is under discussion. For one thing, a curious double standard is used by some corporate interests and elected officials. They are loudly indignant about restrictions that prevent immediate environmental damage and may affect employment. Yet they will routinely justify, if not sanctify, the sacrifice of jobs for the sake of other, and more abstract, goals: short-term corporate economies, for example, or the presumed advantages of enhanced international trade. It's of interest that, as the dispute over wastewater in the Pigeon River raged, Champion International announced that its paper mill was up for sale, with unknown consequences for employees. In such circumstances, "jobs" are not thought of by employers as an absolute. Rather, they represent a benefit to local communities that is sometimes outweighed by other considerations.

Further, few support the idea that you must suffer the effects of pollution so that I can work. The more common understanding is that my employer and I can't pollute or degrade shared resources such as air, water, and forestlands, and neither can you. Once past that point, we can address the welcome topics of economic growth and full employment.

Ecosystem damage kills off jobs as well as creating them. It can leave behind a legacy of economic desolation that may endure far longer than the good times the boomers and boosters prefer to celebrate. People along the Pigeon River complain, and the EPA has documented, that fishing and recreation and the jobs they bring, property values, agriculture, and the ability to attract new industry all suffer because the river is polluted by the paper mill.

Finally, though it is not always easy, there are ways to create new employment within a more sustainable economy and a healthier ecosystem. Clean technology pays off handsomely in jobs at times. As one example, the outrage in Georgia at the turn of the century forced the Copper Basin mining companies to capture as much sulfur dioxide gas as they could, instead of releasing it into the atmosphere. They condensed it into sulfuric acid and sold it. Today, the copper mines have played out. The last of those jobs disappeared when the last mine closed in 1987. But several hundred people are still employed in making sulfuric acid, reclaiming it from materials imported into the basin.

Industrial wastelands like the Copper Basin became are, thankfully, rare in the Blue Ridge. But the effects of air pollution, exotic species, disappearing wildlife habitat, and uncontrolled growth are accumulating. The Blue Ridge will still be green in the future, but it may well have been robbed of much of its beauty, its uniqueness, the splendor of its native plants and animals and their myriad linkages. It also risks some dollars-and-cents pauperization, as the economic opportunities afforded by an unspoiled ecosystem recede.

The technology for conserving electricity and for drastically cutting air pollution from cars and power plants already exists. So do alternatives to slow the use of wood and pulp products. Planning techniques that minimize the impact of population growth are already available. Recovery strategies for endangered species already have a fighting chance for success.

The marketplace alone doesn't provide incentives to adopt these solutions, though, or to invent new ones. Realistically only we and our government can "build in" such incentives.

For now, the costs of harm to the ecosystem, and to us, aren't truly figured into the price of land, electricity, car travel, international trade, or wood and paper products. What we know about the Blue Ridge shows that the damage seems cheaper only until those real costs become apparent.

After that much is known, science and economics take a back seat, and our own values take over—the "Justice Holmes" part of the conversation. Of all the questions in the future of the Blue Ridge, one is most significant and, perhaps, the least predictable: What are people going to do?

Even more crucial than detailed solutions is a general, persistent, and audible public will that action be taken and that promises be kept. If they lack that animating factor, all of our intricate, well-intended policies soon become hollow. They begin to resemble superstitious attempts to ward off evil, like jack-o'lanterns—all face and no guts.

If poll data are any clue (see Solutions 21), people who live in and around the Blue Ridge define themselves as environmentalists without apology. They are willing, by substantial majorities, to sacrifice economic growth and employment to prevent environmental damage. A surprising 40 percent even say they give money to environmental groups—but they are less likely to write to political representatives about environmental problems or to attend meetings.

Americans generally, another poll found, define themselves as environmentalists but have some doubts about whether they can influence the fate of the environment. Some may feel like ecosystem scientists I've talked

SOLUTIONS 21

Public Opinions on Environmental Issues in the Blue Ridge Region

Southern Focus Poll, October 1996

[788 persons, 18 or older]

In general, when you think about the current laws and regulations to protect the environment do they go too far, strike about the right balance, or not go far enough?

Southern Sample:

14.80%	Go too far
26.40%	Strike about the right balance
52.80%	Don't go far enough
6.00%	Don't know/no answer

Carolina Poll, October 1996

[868 North Carolina residents, 18 or older]

Which of these statements comes closer to your own point of view?
1. Protection of the environment should be given priority, even at the risk of curbing economic growth. Or,
2. Economic growth should be given priority even if the environment suffers to some extent.

60.90%	Priority to environment
23.10%	Priority to growth
16.00%	No opinion/can't decide

Carolina Poll, October 1993

[604 North Carolina residents, 18 or older]

Which of these statements comes closer to your own point of view?
1. Protection of the environment should be given priority, even at the risk of curbing economic growth. Or,
2. Economic growth should be given priority even if the environment suffers to some extent.

Responses:

64.40%	Protect environment
26.30%	Economic growth
9.30%	No opinion

Southern Focus Poll, October 1993

[892 persons 18, or older]

Which of these statements comes closer to your own point of view?

1. *Protection of the environment should be given priority, even at the risk of curbing economic growth. Or,*
2. *Economic growth should be given priority even if the environment suffers to some extent.*

    Responses (Southern Sample):

63.80%	Protect environment
21.60%	Economic growth
10.50%	No opinion/can't decide
4.10%	Don't know/no answer

Southern Focus Poll, October 1993

[892 persons, 18 or older]

Do you feel that the environment where you live has become better or worse in the last ten years or has stayed about the same?

    Southern Sample:

12.70%	Better
48.30%	Worse
37.00%	About the same
2.10%	Don't know/no answer

Carolina Poll, October 1993

[604 North Carolina residents 18, or older]

Do you feel that the environment where you live has become better or worse in the last ten years or has stayed about the same?

12.10%	Better
48.60%	Worse
36.60%	About the same
2.70%	Don't know/no answer

Southern Focus Poll, October 1993

[Persons 18 or older]

In general, do you think southerners are more concerned about the environment or less concerned about the environment than people in other areas of the country?

Southern Sample (892 persons):

44.30%	Southerners more
25.90%	Southerners less

SOLUTIONS 21 (cont.)

Southern Sample (892 persons) (continued):

16.70%	About the same (volunteered response)
13.20%	Don't know/no answer

Non-Southern Sample (447 persons):

23.80%	Southerners more
20.30%	Southerners less
27.50%	About the same (volunteered response)
28.40%	Don't know/no answer

South Carolina State Omnibus Survey, April 1992

[770 South Carolina residents, 18 or older]

Many people who live in your area feel that developing the area's economy is important. However, some people feel that increasing the number of jobs in the area should continue even if this means some damage to the environment. Others feel that protecting the environment is more important and that the environment should be protected even if it means the number of jobs would stay the same. Which do you feel is more important . . . increasing the number of jobs in the area or protecting the environment?

40.60%	Increase # jobs
59.40%	Protect environment

South Carolina State Omnibus Survey, April 1992

[779 South Carolina residents, 18 or older]

Do you think it is more important to maintain the water quality in rivers and bays and keep the number of jobs about the same as it is now? Or, to increase the number of jobs but have lower water quality in rivers and bays?

81.00%	Maintain water quality
19.00%	Increase # jobs

Sources: Data archive, Institute for Research in Social Science, University of North Carolina, Chapel Hill. Southern Focus Polls are sponsored by the Center for the Study of the American South and the University of North Carolina's Institute for Research in Social Science. The Carolina Poll is sponsored by the Institute for Research in Social Science and the UNC School of Journalism and Mass Communication. The South Carolina State Omnibus Survey is sponsored by the Survey Research Laboratory, Institute of Public Affairs, University of South Carolina.

with: knowledgeable but powerless. They sometimes describe themselves as "bystanders at a train wreck."

One October night I packed a small telescope out along a ridge near the top of Mount Rogers, the highest peak in Virginia. The wind was gentle but carried a penetrating chill. A row of conifers serrated the twilight on the western horizon. Overhead were the galaxies—immense, eternal, dizzying.

The view, and the vertigo, reminded me of the occasional suggestion that we console ourselves about threats to the ecosystem by taking the longest of long views. Time and space are so vast, we are told, and our influence, after all, is so paltry. Why worry? It's a beguiling, tranquilizing approach. Perhaps "departure" is a better word.

Paleontologist Niles Eldredge, for one, rejects it, even while looking back over the long, turbulent fossil record of extinctions, ice ages, and evolutionary regenerations, and our own modest presence. "This is different," he writes. "This is not just any ecosystem—it is *our* ecosystem, our own very existence as a species, that is at stake. In the fullness of time, no doubt there will be a replacement. . . . But that's later. *This is now.*"

Looking away too fixedly from where we find ourselves just now overlooks not only imminent loss but potential gains. There is in the Blue Ridge, especially, evidence everywhere of people who have valued the natural world and acted to shape its future. They have succeeded in ways we would sorely miss if they had somehow been distracted or discouraged. A handful of examples:

Their work led to the establishment of the Mount Rogers National Recreation Area and recently prevented its quiet expanses from being cut in half by a multilane state highway.

They campaigned doggedly, early in the century, to create Great Smoky Mountains National Park, even after the federal government had thrown up its hands over the complexity of the project. Tens of thousands in Virginia contributed small amounts to purchase the land that made Shenandoah National Park a gift to the nation.

The tiny but potent community around Highlands, North Carolina, fought to have an exquisite chain of waterfalls along the Horsepasture River declared a National Wild and Scenic River and saved them from a dam planned by a hydro-power company.

The citizens of Winchester, Virginia, recently stripped the ribbons and wrapping paper off a proposed glass plant to find a new source of air pollution, and they decided the promise of two hundred new jobs wasn't worth it. The plant won't be built, and the air over nearby Shenandoah National Park will be cleaner as a result.

In Roanoke, developers, planning officials, and local citizens succeeded in minimizing the visual impact of two new housing developments along the Blue Ridge Parkway.

In Tennessee, protected habitat for bears and other wildlife near Great Smoky Mountains National Park is being purchased by the Foothills Land Conservancy. In Georgia, public protests and legal action have forced the Forest Service to begin to account for the effects of clear-cut logging projects on migratory birds in national forests.

An assortment of state agencies in South Carolina, along with nonprofit conservation foundations and plenty of public support, recently joined forces with a landholder, the Duke Energy Corporation, to protect 32,000 acres of wilderness in the exquisite Jocassee Gorge and place them in the public domain. North Carolina may add another 10,000 acres.

Others have pushed, over decades, to fund the research through which we really see much of our ecosystem; for the creation of wilderness areas in the mountains; for clean air and clean water.

These successes have enduring significance for the mountains. But perhaps they also shed some light on our view of the Blue Ridge—a resilient but increasingly fragile national treasure—and our own inescapable role in its future.

Sources are cited below in the order that they have been referred to in the text.

Plate 6

The three images of Plate 6 were derived by Air Resource Specialists, Inc., in the following manner: Speciated aerosol data were collected at Great Smoky Mountains National Park (205 summer sampling days for the period March 5, 1988, to November 29, 1995). Daily data was sorted in ascending order by measured fine mass. Light extinction was reconstructed using the lowest fine mass day and 45 percent relative humidity as the best day, the highest fine mass day and 95 percent relative humidity as the worst day, and the average of all days and 80 percent relative humidity as the average visibility day.

Introduction. Bearings

"Job-killers": Newt Gingrich, U.S. House of Representatives Majority Leader, 1994 and 1995; see Federal News Service, January 25, 1995, "In the News."

Chapter 1. Paupers

thirty million: The figure is conservative. National Park Service counts for its three Blue Ridge units show a total of about twenty-eight million visitors in 1995; visitors to national forests and other Blue Ridge destinations number several million more. National Park Service, "Ten-Year Visitation Report."

nearly three million more: Approximation from 1995 U.S. census data for Blue Ridge counties. U.S. Bureau of the Census, "Projections."

a thousand years: The very general estimates are those of W. Cullen Sherwood, Department of Geology, James Madison University, Harrisonburg, Virginia.

One geomorphologist estimates: Hooke, "On the Efficacy of Humans," 1–5.

the way communities: McNab and Avers, *Ecological Subregions*, 1.

community of rare: Ninety-one different plants are found at this site, as described in Zartman and Pitillo, "Inventory," 50–51.

Blue Ridge extends: Southern Appalachian Man and the Biosphere, *Southern Appalachian Assessment*, 4:145, says there are 10,486,000 acres in the Blue Ridge, or about 16,400 square miles. But this does not include the portion of the "tail" that extends north beyond the Virginia state line. The "Bailey ecoregion" map produced by the U.S. Forest Service is the basis for the boundaries of the Blue Ridge ecosystem used in this book. It shows this northern tail of the Blue Ridge extending to about 40 degrees latitude, into Pennsylvania. My rough estimate is that this tail adds about 700 square miles to the area of the ecosystem.

They are home: Catlin, *Naturalist's Blue Ridge Parkway*, 47, 68, 174–97; Simpson, *Birds*, 278–314; National Park Service, *Handbook*, 53; Linzey, *Mammals*, ix.

130 tree species: Whittaker, "Vegetation," 4.

2-mile-thick glaciers: Delcourt correspondence.

fifty-two peaks: U.S. Geological Survey, *Geographic Names*; Fenneman, *Physiography*, 171.

"If you go anywhere": Jenkins interview.

largest concentration: Southern Appalachian Man and the Biosphere, *Southern Appalachian Assessment*, 4:1.

Visitors to the two: For Great Smoky Mountains National Park tourist spending, see Stynes, *Visitor Spending*, 1–4. Expressed in 1995 Consumer Price Index southern region dollars, using 1995 visitation rates, which yields an estimate of $331 million in direct expenditures. For Shenandoah National Park tourist spending, see Sullivan et al., *Shenandoah National Park*, 1. In 1995 CPI southern region dollars, using 1995 visitation rates to update the study yields an estimate of $45 million in direct expenditures.

"ripple effect": No comparable figures are available for the national forests in the Blue Ridge, which also attract millions of visitors each year.

single year's visitors: Brothers and Chen, *Economic Impact*, 10–11 (in 1992 dollars).

Respondents to a survey: ibid., 21.

Wood products: Southern Appalachian Man and the Biosphere, *Southern Appalachian Assessment*, 4:129, 131.

Blue Ridge itself shelters: ibid., 5:36.

Thirty-six other: ibid., 5:48.

about four dozen varieties: ibid., 2:36–40.

fossil record shows: Jablonski, "Mass Extinctions," 46.

arithmetic of extinction: Pimm et al., "Future of Biodiversity," 347, Pimm interview.

four hundred species are endemic: Nature Conservancy, "National Heritage Central Databases," 1997. This is an incomplete list because so little is known about some species.

our great good fortune: No species endemic to the Blue Ridge has become extinct, so far as is known by the U.S. Fish and Wildlife Service and Nature Conservancy, the most assiduous keepers of the rolls of vanished species. But our knowledge of insects, aquatic species, and animals such as salamanders is limited and recent.

"In most of the other parts of the world": Pimm interview.

By the time: Beattie, "Speech," 5.

"We're nickel-and-diming": Burkhead interview.

major fractions of entire classes: The greatest freshwater mussel diversity in the entire world, about two hundred species, is found in the southeastern United States. Burkhead predicts that 65 to 70 percent of them will vanish from the region or become extinct during the coming century. Burkhead and his colleagues have studied the patterns of both local and regional extinctions among aquatic species in the Southeast. They are the work of humankind. "We've calculated that there is only a one-in-ten-thousand chance that what we're seeing is due to a random natural event," he says (Burkhead interview).

The question . . . has probably occurred to most of us: It first appeared on the national agenda in the late 1970s, when a tiny, rare fish called the snail darter held up construction of the $137 million Tellico Dam in Tennessee, on the outskirts of the Blue Ridge. Biologist Hugh Iltis, the discoverer of rare wild forms of such species as corn and tomatoes that are of great value in improving cultivated food crops, says that the question should be stood on its head: "It should be for them, the sponsors of reckless destruction, to prove to the world that a plant or animal species, or a rare ecosystem, is not useful and not ecologically significant before being permitted by society to destroy it" (Iltis interview).

"There's always been": Sagoff interview.

course of aeons: Leopold, *Round River*, 213.

Not just unknown there: Rock interview.

College students recently found: Hodge et al., "Tolypocladium Inflatum," 715–19.

Two thousand varieties: Houk, *Great Smoky Mountains*, 40.

They consume bits: They eat live blossoms, too, my uncle notes.

worldwide, only four thousand kinds: "No one knows how many different bacteria or fungi there are," adds James Tiedje, director of the Center for Microbial Ecology at the University of Michigan. "Only a small portion have been screened for pharmaceutical products, because we haven't even discovered most of them" (Tiedje correspondence).

"It's very difficult": Zedaker interview.

"the little things": Wilson, "Little Things," 344–46.

Laurel Creek millipede: Hoffman, "Millipeds," 191–92, and Hoffman interview.

"The reality is": Beattie, "Speech," 5.

Settingdown Creek: Freeman interview.

"There's activity": Burkhead interview.

three yardsticks: Tilman et al., "Productivity."

results demonstrate: ibid., 718.

Chapter 2. Balance and Disturbance

last seen for certain: Handley, "Mammals," 556.

seventeen cougar sightings: The year was 1997. Harvey interview.

nearest known population: Brock interview.

hiker from Alexandria: National Park Service, Letter to the Superintendent, 1.

no track: Handley, "Mammals," 556.

Several such "pets": ibid.

"My gut feeling": Harvey interview.

plenty of deer: Brock interview.

Reintroduction efforts . . . low falcon survival rates: Currie interview.

unadapted to the diseases: Catlin, *Naturalist's Blue Ridge Parkway*, 160.

As for red wolves: Defining what a "species" is can be uncertain at times: the red wolves' pedigree as a separate species may never be settled. Tests by one set of geneticists—their findings are disputed by Fish and Wildlife Service science personnel—indicate that red wolves may be a coyote–gray wolf hybrid rather than a separate species.

Even some of those researchers strongly favor the red wolf reintroduction program, however. See Wayne and Gittleman, "Problematic," 38, and Dowling et al., "Genetic Characters," 7–8; Wayne, "Morphologic," 590–92; Nowak, "Hybrid," 593–95; Phillips and Henry, "Comments," 596–99.

last wolf: National Park Service, U.S. Fish and Wildlife Service, "Red Wolf—Recovery," 3.

big wolves: U.S. Senate, "Work of the Biological Survey," map following p. 14, "Distribution of the Big Wolves."

"something we have come to expect": Lucash interview.

"exceedingly dangerous presence": For a fuller record of Senator Helms's remarks, see *Daily Congressional Record*, August 9, 1995, S12014–15; *National Journal's Congress Daily*, August 10, 1995.

104 million cattle: Kate Orr interview.

council recommended: National Research Council, *Endangered Species Act*, 71–72, 96–97, 156–61.

Density in some: McShea interview.

"There is nothing": ibid.

By contrast: ibid.

hungry deer threaten: Caljouw, Caren (see "Unpublished Material" in bibliography).

Deer have browsed: ibid.

"orchids as ice cream": Rock interview.

public opposition: The Park Service's attempts to reduce the numbers of wild ponies it says is damaging vegetation on Shackleford Banks—a part of Cape Lookout National Seashore, North Carolina—is a recent (1997) example. The measures have provoked intense local opposition and a bill in Congress that would take the decision out of the hands of the Park Service.

"It really comes down": Rappole interview.

five days straight: Karish, Blount, and Krumenaker, "Floods and Debris Flows," 1, 36, 37, 42, 43.

Insect, salamander, and fish: ibid., 42, 49.

Such storms: ibid., 6.

"almost every place": Dolloff interview.

Human activities are causing: In the mid-1990s, the rate of worldwide CO_2 emissions was growing by 2 percent annually. The United States emits more CO_2 per person than any other nation, by a wide margin—more than twice as much per person as the Japanese, for example. Marland, Andres, and Boden, "Emissions Estimates."

A few climate scientists: See, for example, *World Climate Report—A Bi-Weekly Report on Global Climate Change*, a newsletter funded by the Western Fuels Association until the spring of 1998, when it listed the Greening Earth Society as its source of funds "with additional funding from other sources." Initial funding for the Greening Earth Society was provided by the Western Fuels Association.

a different conclusion: Watson et al., *Climate Change 1995*.

some general statements: ibid., 6, 26.

Fossilized pollen tells us: Delcourt et al., "History, Evolution, and Organization," 54–55.

Tree species migrate: Delcourt and Delcourt, "Long-Term Forest Dynamics," 309, 326.

took thousands of years: Watson et al., *Climate Change 1995*, 111.

"Entire forest types": ibid., 26, 113.

Fragmentation of forests: Delcourt, Delcourt, and Webb, "Dynamic Plant Ecology," 171. Another unpredictable factor is that green plants respond in different ways when CO_2 concentrations in the atmosphere increase. In the short term, some species grow faster, for example, while others grow more slowly. The long-term effects of added CO_2 on plants are not known. Forests could also experience more frequent outbreaks of disease and insect infestations and more frequent, more intense fires, according to the UN report. See Watson et al., *Climate Change 1995*, 26.

shift the temperature: Watson et al., *Climate Change 1995*, 26.

lop 500 to 1,800 feet: ibid., 26.

Higher or more frequent "spikes": Delcourt correspondence.

"Some species": Watson et al., *Climate Change 1995*, 6.

frog kept swimming: Weiner, *Next One Hundred Years*, 80.

Chapter 3. Vectors

once a broad dominion: Delcourt and Delcourt, "Late Quaternary History," 22.

103 square miles: Dull et al., *Evaluation of Spruce and Fir Mortality*, 1.

more than three-fourths: ibid., 1.

elevations, temperatures: Sharp interview.

Eight plant species: White, "Southern Appalachian Spruce-Fir Ecosystem," 1, 3, 8.

"a place like this": Lovelace, *Mount Mitchell*, 16.

two hundred soldiers: Pyle, "Pre-park Disturbance," 116.

reduced by half: Estimates that these forests were reduced by 90 percent are apparently based on the inclusion of the red spruce highlands of West Virginia, which had no Fraser fir; Saunders, "Recreational Impacts," 100, 102; see also Pyle, "Pre-park Disturbance," 115–30.

eighty years: Eagar interview.

They carried with them: Nicholas and Eagar, "Threatened Ecosystem."

discovered on Mount Mitchell: Eagar, "Biology and Ecology of the Balsam Woolly Aphid," 36.

two to seven years: ibid., 36, 38.

90 percent: Dull et al., *Evaluation of Spruce and Fir Mortality*, 85.

twenty years or less: Cecil Thomas says infestations are already heavy on much younger trees on Mount Rogers (Thomas interview).

seed production in fir trees: Nicholas et al., "Seedling Recruitment," 289–99.

eleven different species: Rabenold et al., "Avian Communities," 3, 4–7.

may have displaced populations: Fies and Pagels, "Northern Flying Squirrel," 584.

it is possible: Fridell interview.

"They are unusual": Harp interview.

joined the endangered species list: *Federal Register*, "Spruce-Fir Moss Spider," 6968–74.

they "dwarf" natural rates: U.S. Congress, *Harmful Non-Indigenous Species*, 78.

great historical convulsions: Elton, *Ecology of Invasions*, 18, 31.

non-native species succeed best: Orians, "Site Characteristics," 133, and U.S. Congress, *Harmful Non-Indigenous Species*, 78, 79.

"biological roulette": U.S. Congress, *Harmful Non-Indigenous Species*, 1.

in human hands: ibid., 78: "Such large-scale movements have become commonplace today, driven by human transformation of natural environments as well as the continual transport of people and cargo around the globe. Resulting rates of species movement dwarf natural rates in comparison."

millions to several billions: ibid., 63, 65.

more frequently: ibid., 96. Rates of introduction of exotic species in this century have been consistently higher than those during the preceding century, and exotic species are "continually being added to the nation's flora and fauna." This study found no evidence of an accelerating rate of new introductions of exotics over the last fifty years, however. Rather, the rate fluctuates widely.

collided three times: Furman correspondence.

last eastward land bridges: Tiffney, "Eocene North Atlantic Land Bridge," 244, 251–59, 265–66; Tiffney adds that the Bering Bridge between North America (Alaska) and Siberia has been a more or less active route of exchange, depending on climate, down to the present.

Humans arrived: Delcourt et al., *Biodiversity*, 62.

twenty-eight kinds of insects: Sailer, "Insect Introduction," 23–26.

more than fifteen hundred: ibid., 23–26.

most recent count: U.S. Congress, *Harmful Non-Indigenous Species*, 3.

known to be harmful: ibid., 79, 301.

Dooley . . . may have acquired . . . specimen may have . . . adelgid was discovered: Miller, "Hemlock Woolly Adelgid," 1–2.

Severe damage, high mortality: Souto, Luther, and Chianese, "Status of HWA," 9–10.

its current rate: ibid., 11.

some of the largest: Langdon and Johnson, "Alien Forest Insects," 10–11.

20 percent of the hemlocks . . . "results and future": Åkerson correspondence.

"There are trees": Tigner interview.

Carolina hemlock may: McClure interview.

twenty million years ago: Wolfe, "Tertiary Floras and Paleoclimates," 51.

When Europeans arrived: Quimby, "Hemlock Ecosystems," 1, 3.

"A hemlock stand": ibid., 3.

trout, too, are more common: ibid., 4.

long-lived and extremely tolerant: ibid., 3.

Three probably depend: Lapin, 1994, cited in Quimby, "Hemlock Ecosystems," 4.

About 14 percent: Neither the Forest Service nor the Park Service has estimated the percentage of hemlocks in Blue Ridge forests; this estimate is from a study of the Smokies by Whittaker, "Vegetation," 38.

Scientists expect: Evans et al., "Potential Impacts," 42–43.

"Aesthetically": Quimby, "Hemlock Ecosystems," 5.

the valuable property: Thomas, "Calcium Accumulation," 115–17.

"We've seen the loss": Robert Anderson interview.

billion chestnut trees: U.S. Congress, *Harmful Non-Indigenous Species*, 66, 74.

a quarter: ibid., 74.

most economically important: U.S. Congress, *Harmful Non-Indigenous Species*, 66.

Five insect species: ibid., 74.

dominant spring leaf-eating insect: Niemela and Mattson, "Invasion," 747.

losses of $764 million: U.S. Congress, *Harmful Non-Indigenous Species*, 67.

when the moth arrived: Åkerson, "Gypsy Moths," 1.

60 percent of the forest: ibid., 1.

Forest Service has predicted: Southern Appalachian Man and the Biosphere, *Southern Appalachian Assessment*, 5:114–17.

That's not to say: Hajek interview.

a few hundred yards: Wallner interview.

up to 60 miles: ibid.

Quotations and details about the Wilmington incident, unless otherwise noted, are from the Bell interview.

"has the potential": U.S. Congress, *Harmful Non-Indigenous Species*, 119.

"There are no guarantees": Wallner interview.

"What we're going to see": Robert Anderson interview.

"So the following day": Rothar interview.

appeared in Brooklyn: Goodman interview.

considered a major pest: Robert Haack, U.S. Forest Service research entomologist, says the Blue Ridge—except, perhaps, for the highest elevations—would be a logical habitat for this species (Haack interview).

"Brooklyn is": Haack et al., "Anoplophora Glabripennis."

decades to centuries: This range of "time to infestation" was suggested by Robert Haack on the low end and Victor Mastro on the high end for spot infestations that might be controlled, rather than total infestation (Haack and Mastro interviews).

"As you can imagine": Haack interview.

More than a thousand: Goodman interview.

results are not certain: ibid.

"Once it gets to the U.S.": Mastro interview.

"When the outrageous": Niemela and Mattson, "Invasion," 751.

"Things get by": Cavey interview.

a daily occurrence: ibid.

The agency's efforts: U.S. Congress, *Harmful Non-Indigenous Species*, 163–231, 287–306.

"Fifty years from now": Haack interview.

severely destructive exotics: The short list includes gypsy moth, ash sawfly, beech scale, pear thrips, and pine shoot beetles.

"A major problem": Robert Anderson interview.

an "ecological disaster": U.S. Congress, *Harmful Non-Indigenous Species*, 74.

decline of birds and turtles: ibid., 74.

16 percent—are exotics: This tally, and the notes that follow, are from Åkerson correspondence.

"invasive and aggressive": National Park Service, Great Smoky Mountains National Park, "Management Plan," app. A, Exotic Plant Category List, 12–13.

"I'm getting more and more worried": Kristine Johnson interview.

"Concerns are increasing": U.S. Congress, *Harmful Non-Indigenous Species*, 74.

Wild hogs: ibid., 73.

Some observers even sense: At least one critic of the National Park Service has made the comparison explicit. See Chase, " 'Nativist' Bias," D-04.

leads . . . to fewer species: "Harmful non-indigenous species also have had profound environmental consequences, exacting a significant toll on U.S. ecosystems. These range from wholesale ecosystem changes and extinction of indigenous species (especially on islands) to more subtle ecological changes and increased biological sameness." U.S. Congress, *Harmful Non-Indigenous Species*, 5. Many exotics "clearly impair biological diversity by causing population declines, species extinctions, or simplification of ecosystems. Moreover, the very establishment of an [exotic] diminishes global biological diversity: as [exotics] . . . spread to more places, these places become more alike biologically." Ibid., 75.

more than 25 percent: ibid., 72.

live white sturgeon: Except as otherwise noted, the details of the white sturgeon story, and Cochran's comments, were supplied by David Cochran in an interview.

their second year: Wydoski and Whitney, *Inland Fishes*, 16–19.

puzzled David Whitmire: Pierson interview.

species is legally sold: In fact, Alabama fisheries administrators purchased two of them from a pet store there and pickled them, so they'd know what the species looks like if the occasion arises, according to Chief of Fisheries Fred Harders (Harders interview).

Conasauga is one of the state's last . . . closely match: Spencer interview.

The state argued: Supreme Court of Georgia, *Department of Natural Resources v. Blue Ridge Mountain Fisheries, Inc., et al.*, 1995; Whitfield Superior Court, *Blue Ridge Mountain Fisheries, Inc., v. Department of Natural Resources*, 1992.

another fisherman: "Know Your Carp," *National Law Journal* 55.

"Maybe in a few more years": Spencer interview.

live for eighty years: Wydoski and Whitney, *Inland Fishes*, 16–19.

20 feet: ibid., 16–19.

largest freshwater fish: ibid., 16–19.

"My immediate interest": Freeman interview.

eat fish eggs . . . carry an assortment of viral diseases: Freeman interview; Spencer interview.

remaining 850: Cochran interview.

into a local stream: Primmer interview.

Chapter 4. Chemical Fates

a million years ago: Behnke interview.

diminished by 70 percent: Moore interview.

Seven of twelve: Species data for the Saint Mary's is from Mohn interview.

Seven out of seventeen: Kauffman, Mohn, and Bugas, "Effects of Acidification." Air pollution and acid precipitation are identified as the "single greatest concern" among many threats to seventy-four insect species the authors identified in Morse, Stark, and McCafferty, "Streams at Risk," 293, 299.

seventy-five thousand to one hundred thousand years: Behnke interview.

"I used to fish": Mohn interview.

a general narrowing: Feldman and Connor, "Influences of Low Alkalinity and pH," 22.

plumes of pollution: The plumes are sulfur dioxide and nitrogen oxides, and what is deposited on the Blue Ridge consists of sulfates and nitrates.

6 percent of the sulfur: Birnbaum, *Feasibility Study*, D-19. This report states that "natural sulfur emissions are estimated to be 6 percent of anthropogenic emissions," which still means that natural emissions are slightly more than 6 percent of the total.

15 percent of the nitrogen oxides: Allen and Gholz, "Impact of Air Pollutants," 85.

rain pH averages 4.3 to 4.5: *Southern Appalachian Assessment*, 3:44; Bruck et al., "Forest Decline," 169.

averages about 4.2: Webb et al., "Acidification," 1367–77.

from 6.8 to about 5: Kauffman, Mohn, and Bugas, "Effects of Acidification."

as low as 4.7: Mohn interview.

pulses of highly acidic water: Birnbaum, *Feasibility Study*, 13.

only real source: ibid., 10.

buffering capacity . . . is declining: Church et al., "Potential Future Effects," 41, 46.

matter of a few decades: ibid., 46.

"The result of these changes": National Park Service, Air Quality Division and Shenandoah National Park, "Adverse Impact Determination for Shenandoah," 22.

North of Roanoke: Webb et al., *Acid-Base Status*, 26–31.

offered a choice: Bulger interview.

"Sublethal" symptoms: ibid.

"In fact, the mix": Bulger interview.

most likely cause: Mohn interview.

Human-made nitrogen emissions: McNulty interview.

fertilization effect: National Park Service, Air Quality Division, "Adverse Impact Determination for Great Smoky Mountains," 8, 9.

nitrogen saturation slows down growth: McNulty interview.

Species diversity, however, *decreased*: Tilman, "Secondary Succession," 189, 213.

highest nitrate concentrations: Stoddard, "Long-Term Changes," cited in Birnbaum, *Feasibility Study*, 12, and Shaver, "Clearing the Air," 694.

"It's that old acidification": Lemley interview.

Thicker soils: Bulger interview.

pollution will remain: Irving, *Acidic Deposition*, vol. 2, report 10, p. 135, and Galloway interview.

"In fact, rarely": Birnbaum, *Feasibility Study*, 16–17.

"Most environmental problems": Kerasote, "Whatever Happened?" 33.

Chapter 5. Pallbearers

Slacks Overlook: near milepost 20.

"forcing many scientists": Shabecoff, "Deadly Combination," 1.

$550 million: King interview, but the estimated cost of National Acid Precipitation Assessment Program varies widely in published reports.

"There is no evidence": Irving, *Acidic Deposition*, vol. 3, report 16, p. 156.

Widespread mortality: ibid., 126

"an initial exploration": ibid., 156, 158.

Is pollution killing trees: "Air Pollution Is Killing Our Trees While the Government Drags Its Feet, in Denial," a headline in an environmental group's newspaper recently charged (see Grant and Flynn, "Killing Our Trees," 12). Several kinds of trees were included in the article's panorama of forest decline in the Southeast, based in part on Forest Service surveys.

But Noel Cost, who supervises those surveys for the Forest Service in the South, explicitly denies signs of impending catastrophe. "Annually, we expect to see mortality in about one percent of the current inventory, and it really hasn't varied that much," he says. "That's just something we haven't observed. We haven't seen anything unusual" (Cost interview).

Cost is not easily dismissed as some sort of apologist for the often controversial

Forest Service, either. He was the coauthor of a report on growth declines in southern pines—they proved to be temporary—that helped kick off concern and research efforts in the 1980s. For the report on growth declines, see Sheffield et al., "Pine Growth Reductions."

On the other hand, as Cost acknowledges, those Forest Serice surveys are not set up to spot more localized instances of forest decline that may occur. They are intended mainly to characterize the growth and condition of forests for their supply of commercial timber, not for ecosystem health.

5,500 feet: the elevation suggested by White in *Spruce-Fir Ecosystem*, 6–10.

acidity of the mists: Bruck, Robarge, and McDaniel, "Forest Decline," 169, and Saxena and Lin, "Cloud Chemistry Measurements," 341, but James Renfro, air resource specialist at Great Smoky Mountains National Park, says Ph can go as low as 2.0 (Renfro correspondence).

90 pounds: Bruck, Robarge, and McDaniel, "Forest Decline," 173.

120 pounds: Saxena and Lin, "Cloud Chemistry Measurements," 329, 345–48, and Saxena interview.

"We've looked at the growth": McLaughlin interview.

rain leaches calcium: McLaughlin et al., "Effects of Acid Deposition," 1–10.

It reduced growth: McLaughlin et al., "Deposition Alters Red Spruce," 380–86.

growth . . . was unaffected: Thornton et al., "Removing Cloudwater," 27.

"crying wolf": Zedaker interview.

nothing unusual going on: The studies of Zedaker and others on spruce growth data are cited in Reams and Van Deusen, "Synchronic Large-Scale Disturbances," 1373.

"Most researchers think": Nicholas interview.

Nearly 70 percent: This figure and those that follow on oak decline are from Oak interview.

complex interaction: Oak decline symptoms were explained by Steven Oak (ibid.).

"probable causal agents": Grant and Loucks, "Epidemiological Assessment," 1.

"There are some plausible explanations": Oak interview.

pine forests, too, may be affected: Forest Service and other surveys in the 1980s reported a decrease in the growth rates of pines in natural upland areas over the preceding thirty years and suggested nine possible causes, one of which was air pollution. See Berrang, Meadows, and Hodges, "Overview," 199, and Zahner, Saucier, and Myers, "Tree-Ring Model," 612–21. A later research study, however, found that there had been no decline. See Berrang, Meadows, and Hodges, "Overview," 199.

exhaustive review: Flagler and Chappelka, "Growth Response," 418–19.

"significant indirect effects": Berrang, Meadows, and Hodges, 203–6.

Hubbard Brook: Likens, Driscoll, and Buso, "Long-Term Effects," 244–46.

aluminum . . . prevents soils: Lawrence, David, and Shortle, "New Mechanism," 162–65.

sugar maples: All data are from Stout interview.

laboratory studies and field observations: Chappelka et al., "Effects of Ozone: Assessment," 1, 4–6; Neufeld et al., "Ozone in Great Smoky Mountains," 594–617, and Chappelka interview.

ninety native plants: Renfro interview.

nine out of ten black cherry trees: National Park Service, Air Quality Division, "Adverse Impact Determination for Great Smoky Mountains," 10

"particular concern": ibid., 7.

NOx in the Southeast: Chameides and Cowling, *Southern Oxidants Study*, ii.

VOCs arise: Human-caused VOC emissions are from automobile engines, gasoline evaporation, wood-burning, petroleum, and chemical industrial sources, evaporation of solvents. Of the natural VOC emissions in the United States, about 90 percent come from trees: 60 percent from conifers, 30 percent from deciduous trees. Allen and Gholz, "Air Quality," 88.

comes from the Midwest: Moy, Dickerson, and Ryan, "Back Trajectories," 2789.

have probably died out: Chappelka interview; Zedaker interview.

"When you lose genetic variations": Chappelka interview.

as yet no proven connections: Skelly interview.

A summary of . . . studies . . . documents: Chappelka et al., "Effects of Ozone: Bibliography."

effect on tree growth is inconclusive: ibid., iii.

adversely affects the growth and health: Flagler and Chappelka, "Growth Response," 418–19.

"an uncomfortable analogy": Wade correspondence.

Hemlocks exposed: McClure, "Nitrogen Fertilization," 227.

strong evidence: Chappelka, "Indirect Effects," 36.

larger fungal disease cankers: Carey and Kelley, "Interaction," 35.

dogwood anthracnose fungus . . . is more virulent: (This does not indicate, however, that dogwood trees would ultimately have fared much better in the absence of acid rain. The fungus would still have done its work, the study's lead author says.) Anderson et al., "Pretreating Dogwood," 55–58, and Robert Anderson interview.

"Concept of multiple stress": Blank, Roberts, and Skeffington, "New Perspectives," 29.

cannot be acknowledged: For a more optimistic view of such research, see Taylor, Johnson, and Andersen, "Air Pollution and Forest Ecosystems," 683.

"Sometimes I wish": Chappelka interview.

"epidemiology of forests": Zedaker interview.

"research about specific outcomes": Loucks interview.

among the foremost skeptics: see Skelly and Innes, "Waldsterben," 1021–32.

"no effects in the parks": Skelly et al., "Panel Discussion," 500.

"On the other hand": Thomson, *Clean Air Partnership*, 92.

Some Forest Service officials say: Noel Cost, for example (Cost interview). He supervises the agency's Forest Inventory and Analysis, often used as ammunition in debates about forest health.

In West Virginia: McKenzie interview.

Also unfortunate . . . "the sample intensity": Bechtold interview.

no data were collected: Burkman interview.

40 to 60 miles: Units used here are miles of "standard visual range," one of three kinds of visibility measurements in common use. The "19 miles" figure is from *Southern Appalachian Assessment*, 3:31. Shining Rock Wilderness visibility data were gathered during the 40 percent of summer days from 1987 to 1993 when clouds and fog did not invalidate visibility measurements. Shining Rock is a few

miles from Sam Knob and the Devil's Courthouse. Forest Service air resource specialist Cindy Huber notes that Shining Rock actually has the best visibility of the "Class I" wilderness viewsheds in the Southern Appalachians: "Most areas are considerably worse," she says (Huber interview).

The lower estimate of the visual range without sulfur pollution is derived from National Research Council and National Academy of Sciences, *Protecting Visibility*, 45, and from conversation with Rudolf Husar, the original source of that data (Husar interview). This very conservative figure for natural or unpolluted visibility is also based on the sulfate concentrations measured on "dirty" summer days at Shenandoah National Park and Great Smoky Mountains National Park March 1988 to February 1994, in Southern Appalachian Man and the Biosphere, *Southern Appalachian Assessment*, 3:34, and a calculation of the amount of light those human-made sulfate concentrations extinguish, courtesy of atmospheric visibility expert Marc Pitchford (Pitchford interview). The impacts of other human-made pollutants, which would have increased the estimate for Shining Rock's natural visibility, were not included. The higher estimate for average natural summer visibility, 60 miles, is from National Park Service, Air Quality Division, "Adverse Impact Determination for Great Smoky Mountains," 23.

The range of natural background visibility figures provided here may well be understated. The estimated annual, as opposed to summer, average natural background visibility range for the whole eastern United States is 93 miles, plus or minus 30 miles. Trijonis, "Natural Background Conditions," 76. Southern Appalachian year-round average visibility now is only about 20 miles—Southern Appalachian Man and the Biosphere, *Southern Appalachian Assessment*, 3: 29–30.

19 miles . . . 8 miles: Huber interview.

70 to 90 percent human-made: Environmental Protection Agency, *Effects of the 1990 Clean Air Act Amendments*, iii—the figure is for "Class I" areas in the rural eastern United States.

"reduces contrast": National Research Council and National Academy of Sciences, *Protecting Visibility*, 1.

number one complaint: National Park Service, Air Quality Division and Shenandoah National Park, "Adverse Impact Determination for Shenandoah," 17.

more than 90 percent: ibid.

10 miles or less: Spitzer correspondence.

view in the Smokies: National Park Service, Air Quality Division, "Adverse Impact Determination for Great Smoky Mountains," 18, 23, 25—data are from 1984 to 1989.

natural sources account for: Environmental Protection Agency, *Effects of the 1990 Clean Air Act Amendments*, iii.

Humidity . . . by itself: Pitchford interview.

On "dirty" summer days: Southern Appalachian Man and the Biosphere, *Southern Appalachian Assessment*, 3:34.

concentrations . . . trended upward: Eldred et al., "Trends," 1; Cahill, Eldred, and Wakabayashi, "Trends," 1–11.

sulfur concentrations were stable: Cahill, Eldred, and Wakabayashi, "Trends," 1, 4, and Eldred interview.

"In 1995, we had": Renfro interview.

during some future July: Watson interview.

penetrate more deeply: Environmental Protection Agency, *Standards for Particulate Matter, OAQPS Staff Paper*, V-3a, V-4.

70 percent: Sisler interview and Sisler et al., *Patterns*, 5-2, 5-3.

clearance . . . takes weeks: Environmental Protection Agency, *Standards for Particulate Matter, Draft Staff Paper*, V-2a.

travel farther down: Environmental Protection Agency, *Standards for Particulate Matter, OAQPS Staff Paper*, V-3.

become more common: ibid., A1.

diseases are also aggravated: ibid., V8–V11.

fifteen thousand a year: Nichols, "Statement by Mary Nichols," April 2, 1997. The figure is "the annual number of premature deaths" that will be avoided if the EPA's latest health standards—"Phase 3," as I have presented it here—are imposed during the first two decades of the new century.

26 percent greater risk: Dockery et al., "Six U.S. Cities," 1753–59.

lives shortened: Lippmann and Thurston, "Health Risk Calculations," 134, and Dockery interview.

"Children are more vulnerable": Bailey and Grupenhoff, "Conference on Air Pollution Impact," 15–16.

"Park visitors don't realize": Renfro interview.

standards have been criticized: Statement of Ronald H. White, deputy director, National Programs, American Lung Association, January 13, 1997: "The American Lung Association has examined the health protection afforded those most vulnerable to particulate matter air pollution by EPA's proposed fine particulate standard, and we find the proposal insufficient. ALA recommends that EPA adopt a daily fine particle standard of no more than 18 micrograms per cubic meter, rather than EPA's proposed level of 50 micrograms per cubic meter" (Ronald White interview). The standards finally adopted turned out to be more lenient: 65 micrograms per cubic meter.

standards were exceeded: data from Sisler correspondence and Copeland interview. Data are estimated from observations sampling one day in three during June, July, and August, on average, during 1993–97.

"Horrendously dirty": Copeland interview.

natural or "background" ozone: Environmental Protection Agency, *Standards for Ozone, Final Staff Paper*, 20–21.

healthy adults suffer: ibid., 23–42. See also Lippmann, "Health Effects of Tropospheric Ozone," 103–23.

"Reactions of some asthmatics": Environmental Protection Agency, *Standards for Ozone, Final Staff Paper*, 58.

"When you get sunburned": MacDonald interview.

ozone "safety limit": The way the new regulation may actually be enforced is much more complex—and more lenient—than just looking for eight-hour averages that exceed 80 ppb, however. You take the fourth-highest, eight-hour ozone averages in an area each year (forget the top three eight-hour averages). Then you average those fourth-highest averages over three years. If that final average exceeds 80ppb, the area will be judged as violating the health standard.

By some measures: James Renfro using the "sum 60" measure (Renfro interview).

sixty-two times: ibid.

Shenandoah National Park's ozone: Spitzer, *Shenandoah Park Resource Newsletter*, 8.

halfway between: White Top data are from Virginia State Department of Environmental Quality. No data on consecutive hours were available, however.

Utility industry estimates: Watkins interview.

about four dollars a year: ibid., expressed in 1993 dollars.

rule of thumb: Birnbaum interview.

estimated the costs: Chestnut, *Human Health Benefits*.

a "premature death": We speak here of a "statistical life," the study's author cautions. This figure is the sum of many people answering a question of this kind: "How much would you pay to avoid an additional 1-in-10,000 risk of premature death?" No one was asked, "How much would you pay to avoid your own premature death?"

yearly benefits: Environmental Protection Agency, "Fact Sheet on Human Health Benefits," 1.

"those estimates": Bolch and Lyons, *Apocalypse Not*, 102.

"the prevention of any": Environmental Protection Agency, *Effects of the 1990 Clean Air Act Amendments*, 1.

"lack of commitment": National Research Council and National Academy of Sciences, *Protecting Visibility*, 11.

Only a fraction: The calculation is based on a conservatively estimated 40- to 60-mile average natural background, standard visual range at the two parks during summer, and the actual average summer view of about 15 miles in 1985. See notes regarding natural visibility, above, and Environmental Protection Agency, *Effects of the 1990 Clean Air Act Amendments*, 48.

would increase both VOCs and NOx: Southern Appalachian Man and the Biosphere, *Southern Appalachian Assessment*, 3:14–17.

One EPA computer model: Thomson, *Clean Air Partnership*, 69.

far above "natural": ibid., 69

1996 summer sulfate: Renfro correspondence.

coming century's sulfur-based pollution: The general outlook is that under Phase 2 rules we would dump as much as a third less sulfate—the by-product of the sulfur dioxide emissions—onto the Blue Ridge each year by 2010. Birnbaum, *Feasibility Study*, xvii.

slowly phased out: ibid., D-18–19.

"Together these factors": National Acid Precipitation Assessment Program, *1990 Integrated Assessment Report*, 1991, 225; see also Birnbaum, *Feasibility Study*, D-19.

EPA projections assume: Birnbaum, *Feasibility Study*, D-19

by the year 2040: ibid., xv, 51.

projections by . . . University of Virginia researchers: Webb et al., *Acid-Base Status*, 67–71, 75; Webb interview.

Phase 2 "means that things": Bulger interview.

timetable for enforcing: Metcalf interview.

Research by a consortium: Northeast States for Coordinated Air Use Management, "Air Pollution Impacts," 1.

rather too forgiving: Soukup, "Statement," 6. See also Environmental Protection Agency, "Regional Haze Regulations," 1.

Chapter 7. Anyplace, U.S.A.

four-story ski condo: The condo is part of the Wintergreen resort.

"really been wide open": Bair interview.

"nearly a complete conversion": Collins interview and Thomas Jefferson Planning district, *Build-Out Analysis*, ix.

The eventual "build-out": Collins interview.

data on population growth: U.S. Bureau of the Census, "Population Distribution and Estimates," 1996.

"phenomenal": Bachtel interview.

A new birth doesn't: ibid.

does not include most seasonal residents: ibid.

"When I moved": Bartlett interview.

on the basis: North Carolina Department of Transportation, *U.S. 19*, 163.

The cycle ends at times in gridlock like that of Gatlinburg, Tennessee, where a wall of commercial development lines the boundary with Great Smoky Mountains National Park. Gatlinburg's waffle joints, T-shirt shops, and faux Bavarian outcroppings may foretell the look of other "gateway" developments in the shrinking green margins around national parks and natural areas.

"roadless" is defined: Southern Appalachian Man and the Biosphere, *Southern Appalachian Assessment*, 4:177, and Romanowski interview.

about 3 percent: Southern Appalachian Man and the Biosphere, *Southern Appalachian Assessment*, 4:178.

more than a third: ibid.

Georgia is mulling over: Ken Anderson interview.

"preferred" route: North Carolina Department of Transportation, *U.S. 19*, p. 5 and fig. 2A.

It may become: Jones interview.

"to ensure that": N.C. Department of Transportation, "Highway Trust Fund," 2.

16-mile section: Wlasha interview.

"The eventual outcome": Gary Everhardt, Blue Ridge Parkway Superintendent, to Pete Sensabaugh, District Construction Engineer, Virginia Department of Transportation, July 11, 1995.

"The solution is": Klein interview.

"I think we're going to": Everhardt interview.

"You have to be": Imhoff, Katherine (see "Unpublished Material" in bibliography).

Chapter 8. Edge

"minor disturbances": North Carolina Department of Transportation, *Highway 19*, 37.

It crosses the Cherokee: Department of the Interior, *Environmental Impact Statement*, 2, Findley interview; McKinney interview.

"This type of unique area": William A. Urgely, acting deputy assistant to the secretary of the interior, to Elmer R. Haile Jr., acting regional engineer, Region 15, Federal Highway Administration, December 5, 1973.

"severe long-term effects": Harvey Bray, director, Tennessee Game and Fish Commission, to Elmer R. Haile Jr., acting regional engineer, Region 15, Federal Highway Administration, November 28, 1973.

"where northern flying squirrels": Fies and Pagels, "Northern Flying Squirrel," 584.

"The road that they put in there": Weigl interview.

Snowbird was not recommended: Findley interview.

"Who are these": Lopez, "Animals We Kill," 19.

species of small mammals: Adams and Geis, "Effects of Roads," 403.

Skunks, weasels, red foxes: North Carolina Department of Transportation, *Highway 19*, 39.

federal estimate: Brown interview.

curtailed legal protection: Gottlieb, *Wise Use Agenda*, 12.

"screw it": Limbaugh, *Ought to Be*, 161.

"All along the Smokies park": Pelton interview.

Bears devour acorns: and other bear behavior is from Pelton interview.

90 percent: McLean and Pelton, "Population Density," 253.

estimated bear count: Pelton interview.

"situation is aggravated": See also Southern Appalachian Man and the Biosphere, *Southern Appalachian Assessment*, 5:87.

"The sad truth": Clark and Pelton, "Management," 15 (draft).

in four hundred cases: Pelton interview.

less than an even chance: ibid.

study of bear behavior: Van Manen and Pelton, "Black Bear Habitat," 323–29.

single small land bridge: Pelton interview.

cameras have documented: Van Manen, Spicer, and Pelton, "Interstate Passageways," 6 (draft).

"70- to 90-kilogram bear": ibid., 7.

"We woke up": Freeman interview.

"Best management practices": The Chattahoochee/Oconee National Forest administration has also complained that private logging and development degrades water quality on and adjacent to national forest lands. See Department of Agriculture, Forest Service, Chattahoochee/Oconee National Forests, *Management Plan*, 47.

energy stored in a leaf: This description was given me by Noel Burkhead (Burkhead interview).

"I've seen private timbering sites": ibid.

Etowah is rich: ibid.

farther downstream: Freeman interview.

"moderate to severe degradation": Southern Appalachian Man and the Biosphere, *Southern Appalachian Assessment*, 2:59.

"The striking thing": Hunter interview.

ovenbirds . . . declined by more than a fourth: Peterjohn interview.

biologists designed an experiment: Marquis and Whelan, "Insectivorous Birds," 2010, 2012, 2013.

"These birds just get walloped": Hunter interview.

all prey on nests: Wilcove, "Nest Predation," 1211–14.

experiment in the Smokies: ibid.

researchers tried to sort it . . . "We suggest": Rich, Dobkin, and Niles, "Forest Fragmentation," 1109–21.

"No matter how carefully": Askins, "Hostile Landscapes," 1956–57.

James and colleagues reanalyzed: James, McCulloch, and Wiedenfeld, "New Approaches," 13.

"That was very surprising": James interview.

Blue Ridge a "hotspot": See also Robinson, "Missing Songbirds," 6, which notes the Smokies as an exception in a generally more mixed continental picture.

"Whenever a man hears it": Henry David Thoreau, cited in Bull and Farrand, *Field Guide*, 666.

"In reality, few rates of change": Peterjohn interview.

"We in government": Droege interview.

"It's interesting": Hunter interview.

John Rappole sees: Rappole interview.

Satellite imagery showed: Dirzo and Garcia, "Rates of Deforestation," 84–90.

"Immediate action": ibid.

"You're talking about": Rappole interview. These first seven quoted sentences and a few others appeared in Nash, "Songbird Connection," 26–27.

Rappole has estimated: Rappole, *Ecology of Migrant Birds*, 155.

Seventeen other such species: ibid., 156.

"Nor is the intensity": ibid.

But it's likely that": Southern Appalachian Man and the Biosphere, *Southern Appalachian Assessment*, 4:34–35: "Increased population density across all counties and development of former farms, forests, and pastures removes habitat for most species of wildlife and fish. Of particular concern recently have been declines of populations and habitat for neotropical songbirds, fox, trout, and many other species. Continuing development pushed by human population growth and greater affluence will, in all likelihood, result in even greater losses of habitat and thus impact even more animal populations in the region.

"More people and the resulting greater amounts of land conversions also impact water quantity, quality, and use. As cities expand, as residential developments are created, and as isolated homes and industries are newly developed, streams of all sizes and qualities are impacted. Development near and along streams, particularly attractive to retirees and tourists, effectively removes or alters riparian vegetation and soils. More roads means more flooding, siltation and introduction of pollutants. Greater numbers of people lead to greater water use and treatment costs and greater interruption of natural cycles. Human development and habitation often occur at higher elevations, resulting in downstream impacts. Some impacts are occurring on public land, and even on designated wilderness. Development at higher elevations also impacts the visual qualities of the region. Some developments can be seen from 50 or more miles away. Scenery in the Southern Appalchians has always been one of the most precious resources of the region. Since the late 1900s, tourists and seasonal residents have used the region as a playground and retirement destination. Urban sprawl, strip industrial and business development, and roading have tremendously changed the scenic character of the area. In unmeasured ways, the tourism potential and economic growth of the region have been impacted."

Native Americans repeatedly burned: Meier, Bratton, and Duffy, "Salamanders," 51–52.

"a potent": Williams, *Americans and Their Forests*, 49.

"during most of": Delcourt and Delcourt, "Pre-Columbian Native American Use of Fire," 1013.

fires were concentrated: ibid. Other research indicates that human-caused fires were rare at higher elevations and in forests such as the dense hardwood coves of the Smokies. See Meier, Bratton, and Duffy, "Salamanders," 51–52.

80 percent . . . burned over: Ayers and Ashe, *Southern Appalachian Forests*, 18; one-fourth had been completely cleared, ibid., 17.

"region has suffered": *Senate Reports*, 60th Cong., 1st sess., vol. 2, "Acquiring National Forests," 5; Yarnell, "Southern Appalachians," 37.

trees grown as crop-wood are: See the description in Southern Appalachian Man and the Biosphere, *Southern Appalachian Assessment*, 5:99.

less than 1 percent to 3 percent: The higher estimate is from figures given in ibid., 5:1 (37 million acres in the Southern Appalachian region) and 5:177 (1.1 million acres of possible, but unverified, old growth). The lower figure is derived from adding known old growth cited on pp. 21 and 26–30 of Davis, *Eastern Old Growth*, and employing again the 37-million-acre total area for the Southern Appalachians. A National Biological Service report classifies old-growth deciduous forest ecosystems in all of the East as "critically endangered." That means 98 percent or more are gone. See Noss, Laroe, and Scott, "Endangered Ecosystems," app. B, p. 50.

It overlooked: U.S. Forest Service wildlife biologist Jesse Overcash, for example, has reviewed old overflight pictures from the 1930s to locate hitherto unmapped old growth in the Jefferson National Forest (Overcash interview).

"Chances are": Rawinski interview.

"In the absence of periodic fire": Southern Appalachian Man and the Biosphere, *Southern Appalachian Assessment*, 5:102.

"As the amount of old growth": Trombulak, "Restoration," 307.

bears also like: Pelton, "Importance of Old Growth," 70.

"Because of the history": ibid., 65, 71.

Young forests developing: Franklin, "Diversity," 169–70.

Dead and dying wood . . . "Dead, standing trees": Department of Agriculture, Forest Service, *Management Plan, George Washington National Forest*, 2-15, 2-16.

1 percent of the canopy: Runkle, "Patterns of Disturbance," 1533, 1542.

gaps . . . occur more frequently: Southern Appalachian Man and the Biosphere, *Southern Appalachian Assessment*, 5:94–95.

low-intensity natural disturbances: Jones, "Virgin Forest," 130–48.

three-fourths of Blue Ridge timberlands: Southern Appalachian Man and the Biosphere, *Southern Appalachian Assessment*, 4:83.

funding . . . negligible: The budget figures, not the opinion, are from Department of Agriculture, Forest Service, "Forest and Rangeland Research FY 1997–FY 1999 Analysis," using the 1998 "enacted" figures, and Toliver interview.

old growth—the 0.5 percent to 3 percent: See earlier note this chapter, at "less than 1 percent to 3 percent," for the basis for these figures.

less than a third: The combined acreage of the two national parks plus the national forest districts within the Blue Ridge ecosystem: approximately 3,311,875 acres. Sources: Tom Wade interview; James Anderson interview; Steven Oak correspondence. The acreage for the whole ecosystem south of the Virginia state line is given in *Southern Appalachian Assessment*, 4:145, as 10,486,000.

150 to 400 years: Trombulak, "Restoration," 312–13.

Chapter 10. Afforestation

2.6 million acres: James Anderson interview for the Cherokee National Forest within the Blue Ridge, and Oak correspondence for the remaining national forest acreages.

"ecosystem services" . . . went with them: Senate Documents, "Report of the Secretary," 16–17.

inauguration, in 1911: Southern Appalachian Man and the Biosphere, *Southern Appalachian Assessment*, 4:13.

first national forest: ibid.

84 percent: ibid., 5:24, 167—the Blue Ridge ecosection is mislabeled in the text on p. 167.

planning guidelines under various federal laws: General Accounting Office Reports, Report No. B-257771.

decades of increases: Department of Agriculture, Forest Service, *Fourth Forest*, 374–429.

public pressure, new worries about the health of forest ecosystems: Department of Agriculture, Forest Service, "Program for Forest and Rangeland Resources" (draft), 3-15; "uncertain political breezes" is my own summary.

mere lip service: for example, Frissell, Nawa, and Noss, "Is There Any Conservation Biology," 461–64.

Forest Service tree disease specialist: McKenzie interview.

a third were still designated: Southern Appalachian Man and the Biosphere, *Southern Appalachian Assessment*, 5:30; table c-17, p. 178.

"suitable for harvest"": Department of Agriculture, Forest Service, *Guidance for Conserving*, 14.

President Bill Clinton: Thomas interview.

"even-aged management": these techniques are defined in Southern Appalachian Man and the Biosphere, *Southern Appalachian Assessment*, v. 5, pp. 99–100.

in 1963: Yarnell, "Southern Appalachians," 57.

about 15 percent: 5:123.

250,000 acres: Stoneking correspondence.

"Clear-cuts look bad": Seiler correspondence.

clear-cuts are antithetical: Trombulak, "Restoration," 312.

persist for centuries: Trombulak cites several studies, such as Meier, Bratton, and Duffy, "Salamander," 52.

the commonest form: Hairston, *Salamander Guilds*, 199; Burton and Likens, "Energy Flow," 1078; Burton and Likens, "Populations and Biomass," 541.

thirty-five different species: Conant and Collins, *Field Guide*, cited in Petranka, Eldridge, and Haley, "Effects," 364.

more "biomass": Hairston, *Salamander Guilds*, 199–200.

Research in and near the Craggy Mountains: Petranka, Eldridge, and Haley, "Effects," 363.

"The fate of salamanders": Ash and Bruce, "Impacts," 300–301.

effect of clear-cutting on wildflowers: Duffy and Meier, "Herbaceous Understories," 196–201.

"If anything": ibid., 199.

"The Effects of Clearcutting": Johnson, Ford, and Hale, 433–35.

Forest Service now predicts: Department of Agriculture, Forest Service, "Program for Forest and Rangeland Resources" (draft), 3-15, 3-37.

Scientists . . . have suggested: Selective cutting that mimics gap phase dynamics is "worthy of investigation." Southern Appalachian Man and the Biosphere, *Southern Appalachian Assessment*, 5:94; also Trombulak, "Restoration," 312.

6,000 miles: Evans correspondence.

research that began in the 1930s: Swift, "Forest Access Roads," 313–24. At times, the Forest Service has failed to document whether its own standards and guidelines and best management practices are being followed and how roads and logging are affecting streams. One example: Cherokee National Forest administrators agreed in 1988, in response to a legal challenge to their Forest Plan, to "prepare, prioritize and maintain a list of roads causing damage" to stream areas, devise solutions, seek funding to fix the problems, and present the list annually for public review. In 1992, an administrative review noted that the obligation was unfulfilled. Five years later, the list still did not exist, according to Charles Lewis of the Cherokee National Forest Staff (Lewis interview). See Department of Agriculture, Forest Service, Cherokee National Forest, *5th Year Review*, 6–8. This administrative review at the Cherokee National Forest found that administrators did not properly document compliance with best management practices and, in logged-off areas, had "failed to collect baseline and project data concerning parameters such as temperature, dissolved oxygen, nitrates, iron, copper, alkalinity and turbidity" to show what conditions the streams were in.

"I'm certain": Van Lear interview.

pollution in the Chattooga: Van Lear, Taylor, and Hansen, "Sedimentation," 20; Van Lear interview.

additional 9 percent: Van Lear, Taylor, and Hansen, "Sedimentation," 9, 20.

"As one of": ibid., 4.

"it may take decades": ibid., 36–37.

Roadless areas account for: Southern Appalachian Man and the Biosphere, *Southern Appalachian Assessment*, 4:178.

federal criteria define "roadless": ibid., 4:177.

More than a third: ibid., 4:178.

planning . . . has been geared: ibid., 4:184.

planned to shrink roadless areas: Department of Agriculture, Forest Service, *Program for Forest and Rangeland Resources*, 1990, table E.8.

New roads would increase: ibid., table E.8.

Cherokee National Forest alone built: Department of Agriculture, Forest Service, *5th Year Review*, 8, 20, and James Anderson interview.

plan for the George Washington: Department of Agriculture, Forest Service, *Management Plan, George Washington National Forest*, 2-17, 2-22.

50 to 80 miles: Department of Agriculture, Forest Service, *Management Plan, George Washington National Forest*, 2-17.

the most recent national long-range plan: Department of Agriculture, Forest Service, "Program for Forest and Rangeland Resources" (draft), app. D-25.

eighteen-month moratorium: Department of Agriculture, Forest Service, "Transportation System."

40 percent of the wood: Southern Appalachian Man and the Biosphere, *Southern Appalachian Assessment*, 4:123—this formulation of the Blue Ridge does not include northern Virginia or northern Georgia counties.

10 to 12 percent: ibid., 4:2.

region's total payroll: ibid., 4:131, multiplied by 0.1 to 0.12—the percentage of timber that comes from the region's national forests.

Adding up . . . direct and indirect . . . effects: ibid., 129, 131. "My gut feeling is that these figures are a little high," Wear notes (Wear correspondence).

meet the demand from private lands: If the demand for timber increases in the future, however, those private supplies, too, may dwindle, economists say.

market for a few types: This information and the quoted material appears in Southern Appalachian Man and the Biosphere, *Southern Appalachian Assessment*, 4:117.

jobs will disappear: Department of Agriculture, Forest Service, *Fourth Forest*, 460–61.

group of six counties: Southern Appalachian Man and the Biosphere, *Southern Appalachian Assessment*, 4:124–25.

If they were closed: ibid., 4:125.

"Large urban areas": ibid., 4:173.

"Public land provides": ibid., 4:172.

"Nature and the outdoors": ibid., 4:173.

Census Bureau projects: U.S. Bureau of the Census, "Projections."

steep upward trend: Patton correspondence. These figures denote "recreational visitor-days," defined as units of 12 hours spent at a national forest. They are for all of the Cherokee, Chattahoochee, Pisgah, and Nantahala National Forests, plus the ranger districts within the Washington, Jefferson, and Sumter National Forests that are on the Blue Ridge.

"opportunities are abundant": ibid., 4:140.

tent camping: National Park Service, "Visitation Report."

southerners who responded: Data in this paragraph from Southern Appalachian Man and the Biosphere, *Southern Appalachian Assessment*, 4:163.

"There was a time": Mohney interview.

logging old growth: Department of Agriculture, Forest Service, *Management Plan, George Washington National Forest*, 2-3 to 2-6.

valley also includes: Croy interview.

As a Forest Service document notes: Department of Agriculture, Forest Service, *Final Environmental Impact Statement*, 1-1.

In 2001: Plunkett interview.

"Politically, it always will be": Mohney correspondence.

Chapter 11. Wild Cards

old mine tunnel: Marshall McClung, of Robbinsville, has located the old tunnel and dug as well into old Forest Service archives, especially the field notes of A. W. Tucker, a consulting mining engineer who roamed these mountains under contract to the Forest Service in 1917 in search of commercially valuable minerals (Mc-Clung interview).

people . . . prosperity . . . destruction: Quinn, "Tennessee's Copper Basin," 140.

A stagnant ocean: Quinn, "Susceptibility," 186.

50-square-mile desert: ibid., 181–82.

"one of the most severe": ibid., 191.

legal struggle was eventually heard: Supreme Court of the United States, *Georgia v. Tennessee Copper Company*.

removed about half: Clark interview.

"We have not run into": Ross interview.

so much nitrogen they approach: Renfro interview.

"If the provisions": Delozier and Humphrey, "State Cancels," A1.

"Unfortunately, the state's approach": Frampton to Sundquist, 1.

included a "bail-out clause": Ross interview.

"beloved scar": Rush interview.

"The forest had been destroyed": James, *Writings*, 842–43.

People along the Pigeon . . . EPA has documented: Clark interview and Clark, "Assessment of Downstream Benefits," 1-35.

40 percent even say: Southern Appalachian Man and the Biosphere, *Southern Appalachian Assessment*, 4:64, 66, 70.

Americans generally: ibid., 4:66.

Why worry?: Easterbrook, *Moment on the Earth*, 49–50, for example.

"This is different": Eldredge, *Miner's Canary*, 229.

People in Winchester: Gordon, "Cardinal Glass Plant," 1-2.

Interviews

All interviews were by the author. Unless otherwise noted, they were conducted by telephone from Richmond, Virginia, during 1995–98.

Åkerson, James (forest ecologist, Shenandoah National Park), correspondence

Anderson, James (planning team leader, Cherokee National Forest)

Anderson, Ken (project engineer, Appalachian Scenic Corridor Study, HDR Engineering, Inc., Atlanta)

Anderson, Robert (plant pathologist, U.S. Forest Service)

Bachtel, Douglas (professor, College of Family and Consumer Sciences, University of Georgia)

Bair, Steve (resource manager, Shenandoah National Park)

Bartlett, Alfred (homeowner, Beech Mountain, N.C.)

Bechtold, William (research forester, U.S. Forest Service)

Behnke, John (professor of fishery biology, Colorado State University)

Bell, Philip (plant protection quarantine officer, Animal and Plant Health Inspection Service, Department of Agriculture)

Birnbaum, Rona (director, Environmental Protection Agency acid rain program)

Bradley, Eldon (Virginia sawmill owner), at Buffalo Creek, Va.

Brock, Rainer H. (professor of wildlife ecology, College of Environmental Science and Forestry, State University of New York at Syracuse)

Brown, Gary (coordinating engineer, Federal Highway Administration)

Bulger, Arthur (fish physiologist, University of Virginia)

Burkhead, Noel (icthyologist, National Biological Survey)

Burkman, William (coordinator, Southern Forest Health Monitoring Program, U.S. Forest Service)

Cavey, Joe (entomologist, biological assessment staff, Animal and Plant Health Inspection Service, Department of Agriculture)

Chappelka, Arthur H. (associate professor of forest biology, School of Forestry, Auburn University)

Clark, Matthew (senior economist, Office of Water, Environmental Protection Agency)

Cochran, David (owner, Blue Ridge Mountain Products, Inc.)

Collins, Michael (senior planner, Thomas Jefferson Planning District, Virginia)

Copeland, Scott (visibility data analyst, National Park Service)

Cost, Noel (project leader, Forest Inventory and Analysis, Southern Region, U.S. Forest Service)

Croy, Steve (forest ecologist, U.S. Forest Service)

Currie, Robert (biologist, U.S. Fish and Wildlife Service), correspondence

DeGrove, John (director, Joint Center for Environmental and Urban Problems, Florida Atlantic University/Florida International University)

Delcourt, Paul (paleoecologist, Department of Ecology and Evolutionary Biology, University of Tennessee at Knoxville)

Dockery, Douglas (epidemiologist, Harvard School of Public Health)

Dolloff, C. Andrew (fishery research biologist, U.S. Forest Service)

Droege, Sam (ornithologist, U.S. Geological Survey, Patuxent Wildlife Research Center; former director, Breeding Bird Survey, U.S. Fish and Wildlife Service)

Eagar, Christopher (ecologist, U.S. Forest Service)

Eldred, Robert (Air Quality Group, Crocker Nuclear Laboratory, University of California, Davis)

Evans, Loren (highway engineer, Engineering Program Management, U.S. Forest Service), correspondence

Findley, Frank (assistant district ranger, Cheoah District, Nantahala National Forest)

Freeman, Byron (assistant research scientist, Institute of Ecology, Museum of Natural History, University of Georgia)

Fridell, John A. (fish and wildlife biologist, U.S. Fish and Wildlife Service, Asheville)

Furman, Tanya (geologist, Environmental Sciences Department, University of Virginia), correspondence

Galloway, James N. (chair, Environmental Sciences Department, University of Virginia)

Goodman, Terry (regional program manager, Animal and Plant Health Inspection Service, Department of Agriculture)

Haack, Robert (research entomologist, U.S. Forest Service)

Hajek, Ann (insect pathologist, Cornell University)

Harders, Fred (chief of fisheries, Alabama Department of Conservation and Natural Resources, Game and Fish Division)

Harp, Joel (arachnologist, University of Tennessee)

Harvey, Chip (ecologist, data manager, and cougar sighting coordinator, Shenandoah National Park)

Henry, Gary (red wolf program coordinator, U.S. Fish and Wildlife Service)

Hoffman, Richard (curator of recent invertebrates, Virginia Museum of Natural History)

Hooke, Roger LeB. (geomorphologist, University of Minnesota)

Huber, Cindy (air resource specialist, U.S. Forest Service)

Hunter, William C. (biologist, U.S. Fish and Wildlife Service)

Husar, Rudolf (director, Center for Air Pollution Impact and Trend Analysis, Washington University, St. Louis)

Iltis, Hugh (botanist emeritus, University of Wisconsin–Madison)

James, Frances C. (Department of Biological Science, Florida State University)

Jenkins, Robert (icthyologist, Roanoke College)

Johnson, Kristine (supervisory natural resource specialist, Great Smoky Mountains National Park), at Great Smoky Mountains National Park

Jones, William (public information officer, North Carolina Department of Transportation)

King, Karen (contractor/administrative researcher, National Acid Precipitation Assessment Program)

Klein, Bill (director of research, American Planning Association)

Lemley, Dennis (research biologist, U.S. Forest Service)

Lewis, Charles (transportation planner, Cherokee National Forest)

Loucks, Orie (ecologist, Miami University of Ohio)

Lucash, Christopher (wildlife biologist, U.S. Fish and Wildlife Service), at Great Smoky Mountains National Park

MacDonald, William (medical officer, Office of Research and Development, Environmental Protection Agency)

MacDonald, William (forest pathologist, West Virginia University)

Mastro, Victor (Animal and Plant Health Inspection Service, Department of Agriculture, center coordinator, Cape Cod, Massachusetts)

McClung, Marshall (local historian and former U.S. Forest Service employee, Robbinsville, North Carolina)

McClure, Mark S. (entomologist and chief scientist, Valley Laboratory, Connecticut Agricultural Experiment Station)

McKenzie, Martin (plant pathologist, U.S. Forest Service)

McKinney, David (Tennessee Department of Wildlife Resources)

McLaughlin, Samuel (plant physiologist, Oak Ridge National Laboratories)

McNulty, Steve (Coweeta Research Station, U.S. Forest Service)

McShea, William (research scientist, Department of Conservation, National Zoological Park, Conservation and Research Center, Front Royal, Virginia)

Metcalf, Linda (management analyst, U.S. Environmental Protection Agency, Air Quality Strategies and Standards Division, Office of Air Quality Planning and Standards)

Mistretta, Paul (pesticide specialist, U.S. Forest Service)

Mohn, Larry (Virginia Deparment of Natural Resources)

Mohney, Sharon (silviculturist, U.S. Forest Service)

Montgomery, Michael (research entomologist, U.S. Forest Service)

Moore, Steve (fisheries biologist, Great Smoky Mountains National Park)

Muncy, Jack (land reclamation project leader, Tennessee Valley Authority)

Murdock, Nora (endangered species biologist, U.S. Fish and Wildlife Service)

Nicholas, Niki (ecologist, Tennessee Valley Authority)

Oak, Steven (plant pathologist, U.S. Forest Service)

Orr, Kate (spokesperson, National Cattlemen's Beef Association, Denver)

Orr, Will (landscape architect, Blue Ridge Parkway, at Asheville

Overcash, Jesse (wildlife biologist, U.S. Forest Service)

Patton, Carol (recreation information manager, U.S. Forest Service, Region 8)

Pelton, Michael (bear biologist, University of Tennessee)

Peterjohn, Bruce (director, Breeding Bird Survey, Biological Resources Division, U.S. Geological Survey)

Pierson, J. Malcolm (senior aquatic biologist, Alabama Power Company)

Pimm, Stuart (ecologist, University of Tennessee)

Pitchford, Marc (staff scientist, Cooperative Institute for Atmospheric Sciences and Terrestrial Applications, Desert Research Institute, Reno, Nevada)

Plunkett, David (forest planner, George Washington National Forest, U.S. Forest Service)

Primmer, Kim (regional fisheries supervisor, Georgia Department of Natural Resources)

Rabenold, Kerry (biologist, Purdue University)

Rappole, John (research coordinator, Department of Conservation, National Zoological Park, Conservation and Research Center, Front Royal, Virginia)

Rawinski, Tom (ecologist, Virginia Department of Conservation and Recreation), at James River Face Wilderness

Renfro, James R. (air resource specialist, Great Smoky Mountains National Park)

Rock, Janet (rare and endangered plant specialist, Great Smoky Mountains National Park), at Great Smoky Mountains National Park

Romanowski, John (U.S. Forest Service)

Ross, Molly (special assistant to the assistant secretary for Fish, Wildlife, and Parks, Department of the Interior)

Rothar, Harry (forestry inspector, Department of Parks and Recreation, Brooklyn, New York)

Rush, Ken (director, Ducktown Basin Museum)

Sagoff, Mark (director, Institute for Philosophy and Public Policy, University of Maryland)

Saxena, V. K. (Department of Marine, Earth, and Atmospheric Sciences, North Carolina State University)

Seiler, John (forest biologist, Virginia Polytechnic and State University), correspondence

Sharp, John (park superintendent, Mount Mitchell State Park)

Sherwood, W. Cullen (Department of Geology, James Madison University)

Sisler, James F. (research coordinator, Cooperative Institute for Research in the Atmosphere)

Skelly, John (phytopathologist, Pennsylvania State University)

Spencer, Michael (senior fisheries biologist, Georgia Department of Natural Resources)

Spitzer, Shane (air quality monitoring specialist, Shenandoah National Park), correspondence

Stein, Bruce (biologist, The Nature Conservancy)

Stoneking, Karl (unit leader, silviculture and genetic resource management, U.S. Forest Service)

Stout, Susan (research project leader, U.S. Forest Service, Northeastern Forest Experiment Station)

Strickland, Wayne (executive director, Fifth Planning District, Virginia)

Thomas, Cecil (wildlife biologist, U.S. Forest Service, Mount Rogers National Recreation Area)

Thomas, Jack Ward (professor of wildlife conservation, School of Forestry, University of Montana; former chief, U.S. Forest Service)

Thompson, Mike (Forest Inventory and Analysis, U.S. Forest Service)

Tiedje, James M. (director, Center for Microbial Ecology, Michigan State University)

Tigner, Tim (forest health specialist, Virginia Department of Forestry)

Toliver, John (research budget coordinator, Office of the Deputy Chief of Forest Service Research, U.S. Forest Service)

Van Lear, David (professor of forestry, Department of Forest Resources, Clemson University)

Wade, Bill (former superintendent, Shenandoah National Park), correspondence

Wade, Tom (statistician, National Park Service, Denver)

Wallner, William (senior research forest entomologist, Northeastern Forest Experiment Station, Northeastern Center for Forest Health Research, U.S. Forest Service)

Watkins, Anne Miller (economist, Acid Rain Division, Environmental Protection Agency)

Watson, John (research scientist, Desert Research Institute, Nevada)

Wear, David (project leader, Economics of Forest Protection and Management, U.S. Forest Service)

Webb, J. R. (coordinator, Shenandoah Watershed Monitoring Project, University of Virginia)

Weigl, Peter (zoologist, Wake Forest University)

White, Peter S. (botanist, University of North Carolina, Chapel Hill)

White, Ronald H. (deputy director, National Programs, American Lung Association)

Wlasha, Butch (Federal Highway Administration, Washington, D.C.)

Zahner, Robert (emeritus professor of forestry, Clemson University), at Highlands, N.C

Zedaker, Shepard (forest ecologist, Virginia Polytechnic Institute and State University)

Books and Articles

Adams, Lowell W., and Aelred D. Geis. "Effects of Roads on Small Mammals." *Journal of Applied Ecology* 20 (1983): 403–15.

Allen, E. R., and H. L. Gholz. "Air Quality and Atmospheric Deposition in Southern U.S. Forests." In Fox and Mickler, *Impact of Air Pollutants*, 83–170.

Anderson, Robert, Paul Berrang, John Knighten, K. Ann Lawton, and Kerry O. Britten. "Pretreating Dogwood Seedlings with Simulated Acidic Precipitation Increases Dogwood Anthracnose Symptoms in Greenhouse-Laboratory Trials." *Canadian Journal of Forestry Research* 23 (1993): 55–58.

Arendt, Randall. *Growing Greener: Putting Conservation into Local Codes*. Media, Pa.: Natural Lands Trust, Inc., Pennsylvania Department of Conservation and Natural Resources et al., 1997.

Ash, Andrew N., and Richard C. Bruce. "Impacts of Timber Harvesting on Salamanders." *Conservation Biology* 8, no. 1 (1994): 300–301.

Askins, Robert A. "Hostile Landscapes and the Decline of Migratory Songbirds." *Science* 267 (March 31, 1995): 1956–57.

Baker, Beth. "How Science Fared under the 104th Congress." *BioScience* 47, no. 1 (1997): 10.

Berrang, Paul, James S. Meadows, and John D. Hodges. "An Overview of Responses of Southern Pines to Airborne Chemical Stresses." In Fox and Mickler, *Impact of Air Pollutants*, 196–243.

Blank, L. W., T. M. Roberts, and R. A. Skeffington. "New Perspectives on Forest Decline." *Nature* 336 (November 3, 1988): 27–30.

Bolch, Ben, and Harold Lyons. *Apocalypse Not: Science, Economics, and Environmentalism*. Washington, D.C.: Cato Institute, 1993.

Bull, John, and John Ferrand Jr. *The Audubon Society Field Guide to North American Birds—Eastern Region*. New York: Knopf, 1977.

Bruck, R. I., W. P. Robarge, and A. McDaniel. "Forest Decline in the Boreal Montane Ecosystems of the Southern Appalachian Mountains." *Water, Air, and Soil Pollution* 48 (1989): 161–80.

Burton, T. M., and G. E. Likens. "Energy Flow and Nutrient Cycling in Salamander Populations in the Hubbard Brook Experimental Forest, New Hampshire." *Ecology* 56 (1975): 1068–80.

———. "Salamander Populations and Biomass in the Hubbard Brook Experimental Forest, New Hampshire." *Copeia*, no. 3 (1975): 541–46.

Carey, W. A., and W. D. Kelley. "Interaction of Ozone Exposure and Fusarium Sub-glutinans on Growth and Disease Development of Loblolly Pine Seedlings." *Environmental Pollution* 84, no. 1 (1994): 35–43.

Catlin, David T. *A Naturalist's Blue Ridge Parkway.* Knoxville: University of Tennessee Press, 1984.

Chameides, W. L., and Ellis B. Cowling. *The State of the Southern Oxidants Study, Policy-Relevant Findings in Ozone Pollution Research, 1988–1994.* Atlanta: Southern Oxidants Study, 1995.

Chappelka, Arthur H. "Indirect Effects and Long-Term Risks of Air Pollution on Eastern North American Temperate Forest Ecosystems: Insect Outbreaks." In "Long-Term Implications of Climate Change and Air Pollution on Forest Ecosystems," progress report of the IUFRO Task Force "Forest, Climate Change, and Air Pollution," edited by Rodolphe Schlaepfer, 36. *International Union of Forestry Research Organization World Series Vol. 4.* Vienna, 1993.

Chase, Alston. "'Nativist' Bias in Our Environmental Policy Leads to Absurd Results." *Denver Post,* July 30, 1995, 2d ed., D-04.

Church, M. Robbins, Paul W. Shaffer, Keith N. Eshleman, and Barry Rochelle. "Potential Future Effects of Current Levels of Sulfur Deposition on Stream Chemistry in the Southern Blue Ridge Mountains." *U.S., Water, Air, and Soil Pollution* 50 (1990): 39–48.

Clark, Joseph D., and Michael R. Pelton. "Management of a Large Carnivore: Black Bear." In Peine, *Ecosystem Management for Sustainability.*

Conant, R., and J. T. Collins. *A Field Guide to Reptiles and Amphibians of Eastern and Central North America.* 3d ed. Boston: Houghton Mifflin, 1991. Cited in Petranka, Eldridge, and Haley, "Effects of Timber Harvesting on Southern Appalachian Salamanders," 364.

Cushman, John H., Jr. "Moratorium on Protecting Species Is Ended." *New York Times,* May 21, 1996, A1.

Davis, Mary Byrd, ed. *Eastern Old Growth Forests: Prospects for Rediscovery and Recovery.* Washington, D.C.: Island Press, 1996.

Delcourt, Hazel R., Paul A. Delcourt, and Thompson Webb III. "Dynamic Plant Ecology: The Spectrum of Vegetational Change in Space and Time." *Quaternary Science Reviews* 1 (1983): 153–75.

Delcourt, Paul A., and Hazel R. Delcourt. "Vegetation Maps for Eastern North America: 40,000 YR BP to the Present." In *Geobotany II,* ed. Robert C. Romans, 123–65. New York: Plenum Publishing Corp., 1981.

——. "Long-Term Forest Dynamics of the Temperate Zone: A Case Study of Late-Quaternary Forests in Eastern North America." *Ecological Studies* 63 (1987).

——. "Pre-Columbian Native American Use of Fire on Southern Appalachian Landscapes." *Conservation Biology* 11, no. 4 (1997): 1010–14.

Delcourt, Paul A., Hazel R. Delcourt, Dan F. Morse, and Phyllis A. Morse. "History, Evolution, and Organization of Vegetation and Human Culture." In *Biodiversity of the Southeastern United States/Lowland Terrestrial Communities,* ed. William H. Martin, Stephen G. Boyce, and Arthur C. Esternacht, 47–78. New York: John Wiley & Sons, 1993.

DeLozier, Stan, and Tom Humphrey. "State Cancels Air Pollution Accord Linked to Smokies; Senate Committee to Examine Board's Action." *Knoxville News-Sentinel,* March 15, 1996, A1.

Dirzo, R., and M. C. Garcia. "Rates of Deforestation in Los Tuxtlas, a Neotropical Area in Southeast Mexico." *Conservation Biology* 6, no. 1 (1992): 84–90.

Dockery, Douglas W., Arden Pope III, Xiping Xu, John D. Spengler, James H. Ware, Martha E. Fay, Benjamin G. Ferris Jr., and Frank E. Speizer. "An Association between Air Pollution and Mortality in Six U.S. Cities." *New England Journal of Medicine* 329, no. 24 (1993): 1753–59.

Dowling, Thomas E., Bruce D. DeMarais, W. L. Minckley, Michael E. Douglas, and Paul C. Marsh. "Use of Genetic Characters in Conservation Biology." *Conservation Biology* 6, no. 1 (1992): 7–8.

Duffy, David Cameron, and Albert J. Meier. "Do Appalachian Herbaceous Understories Ever Recover from Clearcutting?" *Conservation Biology* 6, no. 2 (1992): 196–201.

Easterbrook, Gregg. *A Moment on the Earth: The Coming Age of Environmental Optimism.* New York: Viking, 1995.

Eldred, Robert A., Thomas A. Cahill, William C. Malm, and Marc L. Pitchford. "Trends in Elemental Concentrations of Fine Particles at Remote Sites in the United States." *Atmospheric Environment* 28, no. 5 (1994): 1009–19.

Eldredge, Niles. *The Miner's Canary: Unraveling the Mysteries of Extinction.* Princeton: Princeton University Press, 1991.

Elton, Charles S. *The Ecology of Invasions by Animals and Plants.* New York: John Wiley & Sons, 1958.

Fenneman, Nevin M. *Physiography of the Eastern United States.* New York: McGraw-Hill, 1938.

Fies, Michael L., and John F. Pagels. "Northern Flying Squirrel." In *Virginia's Endangered Species*, ed. Karen Terwilliger, 583–84. Blacksburg, Va.: McDonald and Woodward Publishing Co., 1991.

Flagler, R. B., and A. H. Chappelka. "Growth Response of Southern Pines to Acidic Precipitation and Ozone." In Fox and Mickler, *Impact of Air Pollutants*, 388–424.

Fox, Susan, and Robert Mickler, eds. "Impact of Air Pollutants on Southern Pine Forests." *Ecological Studies* 118 (1996).

Franklin, Jerry R. "Structural and Functional Diversity in Temperate Forests." In *Biodiversity*, ed. E. O. Wilson, 166–75. Washington, D.C.: National Academy Press, 1988.

Frissell, Christopher A., Richard K. Nawa, and Reed Noss. "Is There Any Conservation Biology in 'New Perspectives'?: A Response to Salwasser." *Conservation Biology* 6, no. 3 (1992): 461–64.

Gebhart, Kristi A., and William C. Malm. "Source Apportionment of Particulate Sulfate Concentration at Three National Parks in the Eastern United States." In *Visibility and Fine Particles*, edited by C. V. Mathai, 898–913. Air and Waste Management Association Publication no. TR-17. Pittsburgh: Air and Waste Management Association, 1990.

Gottlieb, Alan M., ed. *The Wise Use Agenda.* Bellevue, Wash.: Free Enterprise Press, 1989.

Grant, William B., and John Flynn. "Air Pollution Is Killing Our Trees While the Government Drags Its Feet, in Denial." *Appalachian Voice*, Sierra Club Highlands Ecoregion Task Force, Boone, N.C. (Winter 1996): 12, 16–17.

Haack, R. A., J. F. Cavey, E. R. Hoebeke, and K. Law. "Anoplophora Glabripennis:

A New Tree-Infesting Exotic Cerambycid Invades New York." *Newsletter of the Michigan Entomological Society* 41, nos. 2–3 (1996): 1–3.

Hairston, N. G., Sr. *Community Ecology and Salamander Guilds*. New York: Cambridge University Press, 1987.

Hajek, Ann E., Richard A. Humber, and Joseph S. Elkinton. "Mysterious Origin of Entomophaga Maimaiga in North America." *American Entomologist* 41 (Spring 1995): 31–42.

Handley, Charles O., Jr. "Mammals." In *Virginia's Endangered Species*, ed. Karen Terwilliger, 539–616. Blacksburg, Va.: McDonald and Woodward Publishing Co., 1991.

Hatcher, Robert D., Jr. "Tectonics of the Southern and Central Appalachian Internides." *Annual Review of Earth Planetary Sciences* (1987): 337–62.

Hodge, K. T., S. B. Krasnoff, and R. A. Humber. "Tolypocladium Inflatum Is the Anamorph of Cordyceps Subsessilis." *Mycologia* 88, no. 5 (1996): 715–19.

Hoffman, Richard L. "Millipeds." In *Virginia's Endangered Species*, edited by Karen Terwilliger, 190–91. Blacksburg, Va.: McDonald and Woodward Publishing Co., 1991.

Hooke, Roger LeB. "On the Efficacy of Humans as Geomorphic Agents." *GSA Today*, Geological Society of America, 4, no. 9 (1994): 1–5.

Houk, Rose. *Great Smoky Mountains National Park*. Boston: Houghton Mifflin, 1993.

Jablonski, David. "Mass Extinctions: New Answers, New Questions." In *The Last Extinction*, edited by Les Kaufman and Kenneth Mallory, 43–61. Cambridge: MIT Press, 1986.

James, Frances C., Charles E. McCulloch, and David A. Wiedenfeld. "New Approaches to the Analysis of Population Trends in Land Birds." *Ecology* 77, no. 1 (1996): 13–27.

James, William. *Writings, 1878–1899*. New York: Library of America, 1992.

Johnson, A. Sydney, William M. Ford, and Philip E. Hale. "The Effects of Clearcutting on Herbaceous Understories Are Still Not Fully Known." *Conservation Biology* 7, no. 2 (1993): 433–35.

Jones, E. W. "The Structure and Reproduction of the Virgin Forest of the North Temperate Zone." *New Phytologist* 44 (1945): 130–48.

Jones, Gareth, Alan Robertson, Jean Forbes, and Graham Hollier. *HarperCollins Dictionary of Environmental Science*. New York: HarperCollins, 1992.

Kerasote, Ted. "Whatever Happened to Acid Rain?" *Sports Afield* 212 (November 1994): 5.

"Know Your Carp." *National Law Journal*, December 14, 1992, 55.

Langdon, Keith R., and Kristine D. Johnson. "Alien Forest Insects and Diseases in Eastern U.S. National Park Units: Impacts and Interventions." *George Wright Forum* 9, no. 1 (1992): 2.

Lawrence, Gregory B., Mark B. David, and Walter C. Shortle. "A New Mechanism for Calcium Loss in Forest-Floor Soils." *Nature* 378, no. 9 (1995): 162–65.

Leopold, Aldo. *Round River*. Edited by Luna Leopold. 2d ed. Minneapolis: NorthWord Press, 1991.

Likens, G. E., C. T. Driscoll, and D. C. Buso. "Long-Term Effects of Acid Rain: Response and Recovery of a Forest Ecosystem." *Science* 272 (April 12, 1996): 244–46.

Limbaugh, R. *The Way Things Ought to Be*. New York: Simon and Schuster, 1993.

Linzey, Donald W. *Mammals of Great Smoky Mountains National Park*. Blacksburg, Va.: MacDonald and Woodward Publishing Co., 1995.

Lippmann, Morton. "Health Effects of Tropospheric Ozone: Review of Recent Research Findings and Their Implications for Ambient Air Quality Standards." *Journal of Exposure Analysis and Environmental Epidemiology* 3, no. 1 (1993): 103–29.

Lippmann, Morton, and G. D. Thurston. "Sulfate Concentrations as an Indicator of Ambient Particulate Matter Pollution for Health Risk Calculations." *Journal of Exposure Analysis and Environmental Epidemiology* 6, no. 2 (1996): 123–46.

Lopez, Barry. "Who Are These Animals We Kill?" *Harpers* (July 1990): 19–21.

Lovelace, Jeff. *Mount Mitchell, Its Railroad, and Toll Road*. Johnson City, Tenn.: Overmountain Press, 1994.

Marquis, Robert J., and Christopher J. Whelan. "Insectivorous Birds Increase Growth of White Oak through Consumption of Leaf-Chewing Insects." *Ecology* 75, no. 7 (1994): 2007–14.

McClure, Mark. "Nitrogen Fertilization of Hemlock Increases Susceptibility to Hemlock Woolly Adelgid." *Journal of Arboriculture* 17 (1991): 8.

McLaughlin, Samuel B., J. D. Joslin, A. Stone, R. Wimmer, and S. Wullschleger. "Effects of Acid Deposition on Calcium Nutrition and Health of Southern Appalachian Spruce-Fir Forests." In *Proceedings, IUFRO Symposium, Air Pollution and Multiple Stresses*, 1–11. Fredericton, New Brunswick: International Union of Forestry Research Organizations, in press.

McLaughlin, Samuel B., Mark G. Tjoelker, and W. K. Roy. "Acid Deposition Alters Red Spruce Physiology: Laboratory Studies Support Field Observations." *Canadian Journal of Forest Research* 23 (1993): 380–86.

McLean, Peter K., and Michael R. Pelton. "Estimation of Population Density and Growth of Black Bears in the Smoky Mountains." *International Conference on Bear Research and Management* 9, no. 1 (1994): 253–61.

Meier, Albert J., Susan Power Bratton, and David Cameron Duffy. "Biodiversity in the Herbaceous Layer and Salamanders in Appalachian Primary Forests." In Davis, *Eastern Old Growth Forests*, 49–64.

Morse, J. C., B. P. Stark, and W. P. McCafferty. "Southern Appalachian Streams at Risk: Implications for Mayflies, Stoneflies, Caddisflies, and Other Aquatic Biota." In *Aquatic Conservation: Marine and Freshwater Ecosystems* 3, no. 4 (1993): 293–303.

Moy, Leslie A., Russell R. Dickerson, and William F. Ryan. "Relationship between Back Trajectories and Tropospheric Trace Gas Concentrations in Rural Virginia." In *Atmospheric Environment* 28, no. 17 (1994): 2789–2800.

Nash, Steve. "The Songbird Connection." *National Parks* (November–December 1990): 22–27.

Neufeld, H. S., J. R. Renfro, W. D. Hacker, and D. Silsbee. "Ozone in Great Smoky Mountains National Park: Dynamics and Effects on Plants." In *Tropospheric Ozone and the Environment*, vol. 2, *Effects, Modeling, and Control and the Response of Southern Commercial Forests to Air Pollution*, ed. R. Berglund, 594–617. Pittsburgh: Air and Waste Management Association, 1992.

Nicholas, N. S., and Christopher Eagar. "Threatened Ecosystem: High-Elevation Spruce-Fir Forest." In Peine, *Ecosystem Management for Sustainability*.

Nicholas, N. S., S. M. Zedaker, C. Eagar, and F. T. Bonner. "Seedling Recruitment and Stand Regeneration in Spruce-Fir Forests of the Great Smoky Mountains." *Bulletin of the Torrey Botanical Club* 119, no. 3 (1992): 289–99.

Niemela, Pekka, and William J. Mattson. "Invasion of North American Forests by European Phytophagous Insects." *BioScience* 46 (November 1996): 741–53.

Nowak, Ronald M. "The Red Wolf Is Not a Hybrid." *Conservation Biology* 6, no. 4 (1992): 593–95.

Orians, G. H. "Site Characteristics Favoring Invasions." In "Ecology of Biological Invasions of North America and Hawaii," edited by H. A. Mooney and J. A. Drake, *Ecological Studies* 58 (1986): 133–48.

Peine, J., ed. *Ecosystem Management for Sustainability: Principles and Practices as Illustrated by a Regional Biosphere Cooperative* [working title]. Delray Beach, Fla.: St. Lucie Press, in press.

Pelton, Michael R. "The Importance of Old Growth to Carnivores in Eastern Deciduous Forests." In Davis, *Eastern Old Growth Forests*, 65–75.

Petranka, James W., Matthew E. Eldridge, and Katherine E. Haley. "Effects of Timber Harvesting on Southern Appalachian Salamanders." *Conservation Biology* 7, no. 2 (1993): 363–70.

Phillips, Michael K., and Gary V. Henry. "Comments on Red Wolf Taxonomy." *Conservation Biology* 6, no. 4 (1992): 596–99.

Pimm, Stuart L., Gareth J. Russell, John L. Gittleman, and Thomas M. Brooks. "The Future of Biodiversity." *Science* 269 (July 21, 1995): 347–50.

Quimby, John W. "Value and Importance of Hemlock Ecosystems in the Eastern United States." In Salom, Tigner, and Reardon, *Proceedings*, 1–8.

Quinn, M.-L. "Tennessee's Copper Basin: A Case for Preserving an Abused Landscape." *Journal of Soil and Water Conservation* 43, no. 2 (1988): 140–44.

———. "The Appalachian Mountains Copper Basin and the Concept of Environmental Susceptibility." *Environmental Management* 15, no. 2 (1991): 179–94.

Rabenold, Kerry N., Peter T. Fauth, Bradley W. Goodner, Jennifer A. Sadowski, and Patricia T. Parker. "Response of Avian Communities to Disturbance by an Exotic Insect in Spruce-Fir Forests of the Southern Appalachians." *Conservation Biology*, in press.

Rappole, John H. *The Ecology of Migrant Birds: A Neotropical Perspective*. Washington, D.C.: Smithsonian Institution Press, 1995.

Reams, Gregory A., and Paul C. Van Deusen. "Synchronic Large-Scale Disturbances and Red Spruce Growth Decline." *Canadian Journal of Forest Research* 23 (1993): 1361–74.

Rich, Adam C., David S. Dobkin, and Lawrence J. Niles. "Defining Forest Fragmentation by Corridor Width: The Influence of Narrow Forest-Dividing Corridors on Forest-Nesting Birds in Southern New Jersey." *Conservation Biology* 8, no. 4 (1994): 1109–21.

Robinson, Scott K. "The Case of the Missing Songbirds." *Consequences* 3 (1997): 3–16.

Robinson, Scott K., Frank R. Thompson III, Therese M. Donovan, Donald R. Whitehead, and John Faaborg. "Regional Forest Fragmentation and the Nesting Success of Migratory Birds." *Science* 267 (March 31, 1995): 1987–90.

Runkle, James Reade. "Patterns of Disturbance in Some Old-Growth Mesic Forests of Eastern North America." *Ecology* 63, no. 5 (1982): 1533–46.

Sailer, R. I. "History of Insect Introduction." In *Exotic Plant Pests and North Ameri-*

can Agriculture, ed. C. Graham and C. Wilson, 15–38. New York: Academic Press, 1983.

Sauer, J. R., J. E. Hines, G. Gough, I. Thomas, and B. G. Peterjohn. *The North American Breeding Bird Survey Results and Analysis, 1996*. Laurel, Md.: Patuxent Wildlife Research Center.

Saxena, V. K., and N.-H. Lin. "Cloud Chemistry Measurements and Estimates of Acidic Deposition on an Above Cloudbase Coniferous Forest." *Atmospheric Environment* 24A, no. 2 (1990): 329–52.

"Senate Interior Bill OKd; Votes Stacked on Transportation." *National Journal's Congress Daily*, August 10, 1995.

Shabecoff, Phillip. "Deadly Combination Felling Trees in East." *New York Times*, July 24, 1988, 1.

Shaver, Christine L., Kathy A. Tonnessen, and Tonnie G. Maniero. "Clearing the Air at Great Smoky Mountains National Park." *Ecological Applications* 4, no. 4 (1994): 690–701.

Shostak, Arthur. "The Nature of Work in the Twenty-first Century: Certain Uncertainties." *Business Horizons* 36, no. 6 (1993): 30.

Simberloff, Daniel. "Introduced Insects: A Biogeographic and Systematic Perspective." In Mooney and Drake, *Ecology of Biological Invasions of North America and Hawaii*, 3–26.

Simpson, Marcus B., Jr. *Birds of the Blue Ridge Mountains*. Chapel Hill: University of North Carolina Press, 1992.

Skelly, John M., and John L. Innes. "Waldsterben in the Forests of Central Europe and Eastern North America: Fantasy or Reality?" *Plant Disease* 78, no. 11 (1994): 1021–32.

Skelly, John M., Allen S. Lefohn, Richard B. Flagler, Joseph E. Miller, and Beverley Hale. "A Panel Discussion: The Identification of Ambient Levels of Ozone That Will Protect Vegetation." In *Proceedings of a U.S. EPA/Air and Waste Management Association International Specialty Conference, Tropospheric Ozone: Non-Attainment and Design Value Issues*, ed. Jaroslav J. Vostal, 493–509. Pittsburgh: Air and Waste Management Association, 1992.

Souto, Dennis, Tom Luther, and Bob Chianese. "Past and Current Status of HWA in Eastern and Carolina Hemlock Stands." In Salom, Tigner, and Reardon, *Proceedings*, 9–15.

Stoddard, J. L. "Long-Term Changes in Watershed Retention of Nitrogen: Its Causes and Aquatic Consequences." In *Environmental Chemistry of Lakes and Reservoirs*, ed. L. S. Baker, 223–84. Advances in Chemistry 237. Washington, D.C.: American Chemical Society, 1994.

Swift, L. W., Jr. "Forest Access Roads: Design, Maintenance, and Soil Loss." In "Forest Hydrology and Ecology at Coweeta," edited by Wayne T. Swank and D. A. Crossley Jr. *Ecological Studies* 66 (1987): 313–24.

Taylor, George E., Dale W. Johnson, and Christian P. Andersen. "Air Pollution and Forest Ecosystems: A Regional to Global Perspective." *Ecological Applications* 4, no. 4 (1994): 662–89.

Teskey, R. O. "Synthesis and Conclusions from Studies of Southern Commercial Pines." In Fox and Mickler, *Impact of Air Pollutants*, 467–90.

Thomas, William A. "Calcium Accumulation and Cycling by Dogwood." *Ecological Monographs* 39, no. 2 (1969): 101–20.

Thornton, F. C., C. McDuffie Jr., P. A. Pier, and R. C. Wilkinson. "The Effects of Removing Cloudwater and Lowering Ambient O_3 on Red Spruce Grown at High
Elevations in the Southern Appalachians." *Environmental Pollution* 79 (1993):
21–29.

Tiffney, Bruce H. "The Eocene North Atlantic Land Bridge and Its Importance in
Tertiary and Modern Phytogeography of the Northern Hemisphere." *Journal of
the Arnold Arboretum* 66 (1985): 243–73.

Tilman, David. "Secondary Succession and the Pattern of Plant Dominance along
Experimental Nitrogen Gradients." *Ecological Monographs* 57, no. 3 (1987):
189–214.

———. "The Benefits of Natural Disasters." *Science* 273 (September 13, 1996): 1518.

Tilman, David, David Wedin, and Johannes Knops. "Productivity and Sustainability Influenced by Biodiversity in Grassland Ecosystem." *Nature* 379, no.
6567 (1996): 718–20.

Trombulak, Stephen C. "The Restoration of Old Growth: Why and How." In Davis,
Eastern Old Growth Forests, 305–20.

Van Manen, Frank T., and Michael R. Pelton. "Data-Based Modelling of Black Bear
Habitat Using GIS." *Proceedings, International Union of Game Biologists, XXI Congress*, edited by Ian Thompson, 1 (1993): 323–29.

Wagner, Frederic H., Ronald Foresta, R. Bruce Gill, Dale R. McCullough, Michael
R. Pelton, William F. Porter, and Hal Salwasser. *Wildlife Policies in the U.S. National Parks*. Washington, D.C.: Island Press, 1995.

Wallner, W. E. "Invasive Pests ('Biological Pollutants') and U.S. Forests: Whose
Problem, Who Pays?" *European Plant Protection Organization Bulletin* 26 (1996):
167–80.

Watson, Robert T., Marufu Zinyowera, Richard Moss, and David Dokken, eds. *Cli-
mate Change 1995: Impacts, Adaptations, and Mitigation of Climate Change:
Scientific-Technical Analyses, The IPCC Second Assessment Report, Working Group II
Summary for Policymakers*. New York: Cambridge University Press, 1995.

Wayne, Robert K. "On the Use of Morphologic and Molecular Genetic Characters
to Investigate Species Status." *Conservation Biology* 6, no. 4 (1992): 590–92.

Wayne, Robert K., and John L. Gittleman. "The Problematic Red Wolf." *Scientific
American* 273, no. 1 (July 1995): 36–39.

Webb, J. R., B. J. Cosby, J. N. Galloway, and G. M. Hornberger. "Acidification of
Native Brook Trout Streams in Virginia." *Water Resources Research* 25, no. 6
(1989): 1367–77, cited in National Park Service, Air Quality Division, *Technical
Support Document Regarding Adverse Impact Determination for Great Smoky Mountains National Park*, 9.

Weiner, Jonathan. *The Next One Hundred Years: Shaping the Fate of Our Living Earth*.
New York: Bantam Books, 1990.

Whittaker, R. "Vegetation of the Great Smoky Mountains." *Ecological Monographs*
26, no. 1 (1956): 1–80.

Wilcove, David S. "Nest Predation in Forest Tracts and the Decline of Migratory
Songbirds." *Ecology* 66, no. 4 (1985): 1211–14.

Williams, Michael. *Americans and Their Forests: A Historical Geography*. New York:
Cambridge University Press, 1989.

Wilson, E. O. "The Little Things That Run the World." *Conservation Biology* 1
(1987): 344–46.

Wolfe, J. A. "Tertiary Floras and Paleoclimates of the Northern Hemisphere." In

Land Plants: Notes for a Short Course, organized by R. A. Gastaldo and edited by
T. W. Broadhead, 182–96. Department of Geological Sciences, Studies in Geol-
ogy, 15. Knoxville: University of Tennessee, 1985, as cited in Delcourt et al.,
Biodiversity.

Wydoski, Richard S., and Richard R. Whitney. *Inland Fishes of Washington*. Seattle:
University of Washington Press, 1979.

Zahner, Robert, Joseph R. Saucier, and Richard K. Myers. "Tree-Ring Model Inter-
prets Growth Decline in Natural Stands of Loblolly Pine in the Southeastern
United States." *Canadian Journal of Forest Research* 19 (1989): 612–21.

Unpublished Material and Government Documents

Åkerson, James. "Resource Impacts from Gypsy Moths at Shenandoah National
Park, Virginia." Center for Resources, National Park Service, 1997.

Ayers, H. B., and W. W. Ashe. *Forests and Forest Conditions in the Southern Appalachians, in A Report of the Secretary of Agriculture in Relation to the Forests, Rivers,
and Mountains of the Southern Appalachian Region*. Washington, D.C.: Government Printing Office, 1901.

———. *The Southern Appalachian Forests*. 58th Cong., 3d sess., House of Representatives, Doc. No. 409, Professional Paper No. 37, Department of the Interior, U.S.
Geological Survey. Washington, D.C.: Government Printing Office, 1905.

Bailey, Byron J., and John T. Grupenhoff, eds. "Summary, National Conference on
Air Pollution Impact on Body Organs and Systems." National Association of
Physicians for the Environment, Washington, D.C., 1995.

BankAmerica Corporation. "Beyond Sprawl: New Patterns of Growth to Fit the
New California." 1996.

Beattie, Mollie. "Speech to the Society of Environmental Journalists." Los Angeles,
May 20, 1995.

Birnbaum, Rona, ed. *Acid Deposition Standard Feasibility Study, Report to Congress*.
Environmental Protection Agency, Office of Air and Radiation, Acid Rain Division, 1995.

Bray, Harvey. Tennessee Fish and Game Commission. Letter, November 28, 1973,
in *Final Environmental Impact Statement, Tennessee and North Carolina, Tellico
Plains–Robbinsville Highway*, Doc. FHWA-TN-EIS-72-16-F and 72-23-F, August
2, 1977.

Brothers, Gene, and Rachel J. C. Chen. *Economic Impact of Travel to the Blue Ridge
Parkway, Virginia and North Carolina*. Department of Parks, Recreation, and
Tourism Management, North Carolina State University, 1997.

Cahill, Thomas A., Robert A. Eldred, and Paul H. Wakabayashi. "Trends in Fine
Particle Concentrations at Great Smoky Mountains National Park." Paper pre-
sented at the annual meeting of the Air and Waste Management Association,
Nashville, Tenn., June 1996, 1–11.

Caljouw, Caren (natural areas stewardship manager, Virginia Department of Conservation and Recreation). Comments in a symposium at Shenandoah National
Park, May 1997.

Chappelka, Arthur H., Lisa J. Samuelson, John M. Skelly, and Allen S. Lefohn. "Effects of Ozone on Vegetation in the Southern Appalachians: An Assessment of
the Current State of Knowledge." 1996.

Chappelka, Arthur H., Lisa J. Samuelson, John M. Skelly, Allen S. Lefohn, Daureen

Nesdill, and Elizabeth S. Hildebrand. "Effects of Ozone on Vegetation in the Southern Appalachians: An Annotated Bibliography, Report for the Southern Appalachian Mountain Initiative." 1996.

Chestnut, Lauraine. *Human Health Benefits from Sulfate Reductions Under Title IV of the 1990 Clean Air Act Amendments—Final Report to the U.S. Environmental Protection Agency, Office of Air and Radiation, Office of Atmospheric Programs, Acid Rain Division.* Hagler Bailly Consulting, November 1995.

Clark, Matthew. Environmental Protection Agency, Office of Water. "Assessment of Downstream Benefits as a Result of Water Quality Improvements on the Pigeon River." 1997.

Daily Congressional Record, Wednesday, August 9, 1995, vol. 141, #133, 104th Cong., 1st sess, S12014–15.

Delcourt, Hazel R., and Paul A. Delcourt. "Late-Quaternary History of the Spruce-Fir Ecosystem in the Southern Appalachian Mountain Region." In White, *Spruce-Fir Ecosystem*, 22–35.

Department of Agriculture, Forest Service. *The South's Fourth Forest: Alternatives for the Future.* Forest Resource Report No. 24, 1988.

Department of Agriculture, Forest Service. *Forest Service Program for Forest and Rangeland Resources: A Long-Term Strategic Plan.* May 1990.

Department of Agriculture, Forest Service. "The Forest Service Program for Forest and Rangeland Resources: A Long-Term Strategic Plan." Draft. 1995.

Department of Agriculture, Forest Service. "36 CFR Part 212, Administration of the Forest Development Transportation System, Advance Notice of Proposed Rulemaking." 1998.

Department of Agriculture, Forest Service, Chattahoochee/Oconee National Forests. *Forest Land and Resource Management Plan, Five Year Review and Recommendations.* 1993.

Department of Agriculture, Forest Service, Cherokee National Forest. *Land and Resource Management Plan, 5th Year Review.* Doc. #1920-2-1 RA 9611-1, 1993.

Department of Agriculture, Forest Service, Office of the Deputy Chief of Forest Service Research. "Forest and Rangeland Research FY 1997–FY 1999 Analysis." 1998.

Department of Agriculture, Forest Service, Southern Region. *Final Environmental Impact Statement for the Revised Land and Resource Management Plan, George Washington National Forest*, 1993.

Department of Agriculture, Forest Service, Southern Region. *Final Revised Land and Resource Management Plan, George Washington National Forest*, 1993.

Department of Agriculture, Forest Service, Southern Region. *Guidance for Conserving and Restoring Old-Growth Forest Communities on National Forests in the Southern Region.* Forestry Report R8-FR 62, 1997.

Department of the Interior. Letter dated December 5, 1973, in *Final Environmental Impact Statement, Tennessee and North Carolina, Tellico Plains–Robbinsville Highway.* Doc. FHWA-TN-EIS-72-16-F and 72-23-F, August 2, 1977.

Dull, C. W., J. E. Ward, H. D. Brown, G. W. Ryan, W. H. Clerke, and R. J. Uhler. *Evaluation of Spruce and Fir Mortality in the Southern Appalachian Mountains.* Department of Agriculture, Forest Service, Protection Report R8-PR 13, Atlanta, 1988.

Eagar, Christopher. "Review of the Biology and Ecology of the Balsam Woolly Aphid in Southern Appalachian Spruce-Fir Forest." In White, *Spruce-Fir Eco-system*, 36–50.

English, D. B. K., Carter Betz, J. M. Young, John C. Bergstrom, and H. Ken Cordell. "Outdoor Recreation and Wilderness Assessment Research." Department of Agriculture, Forest Service, Athens, Ga., 1990, in *Regional Demand and Supply Projections for Outdoor Recreation*, General Technical Report RM-230, Department of Agriculture, Forest Service, Rocky Mountain Forest and Range Experiment Station, Ft. Collins, August 1993.

Environmental Protection Agency. *Review of the National Ambient Air Quality Standards for Particulate Matter: Policy Assessment of Scientific and Technical Information, Draft Staff Paper*, November 1995.

——. *Review of the National Ambient Air Quality Standards for Particulate Matter: Policy Assessment of Scientific and Technical Information, OAQPS Staff Paper*. July 1996.

Environmental Protection Agency, Office of Air and Radiation, Office of Atmospheric Programs, Acid Rain Division. "Fact Sheet on Human Health Benefits from Sulfate Reductions under Title IV of the 1990 Clean Air Act Amendments—Final Report." December 1995.

Environmental Protection Agency, Office of Air Quality Planning and Standards. *Effects of the 1990 Clean Air Act Amendments on Visibility in Class I Areas: An EPA Report to Congress*. 1993.

——. *Review of National Ambient Air Quality Standards for Ozone—Assessment of Scientific and Technical Information, Draft Staff Paper*. 1995.

——. *Review of National Ambient Air Quality Standards for Ozone—Assessment of Scientific and Technical Information, Final Staff Paper*. June 1996.

Environmental Protection Agency, Air Quality Strategies and Standards Division, Visibility and Ecosystem Protection Group. "Fact Sheet—Proposed Regional Haze Regulations for Protection of Visibility in National Parks and Recreation Areas." July 18, 1997.

Evans, Richard A., Elizabeth Johnson, Jeff Shreiner, Allan Ambler, John Battles, Natalie Cleavitt, Tim Fahey, Jim Sciascia, and Ellen Pehek. "Potential Impacts of Hemlock Woolly Adelgid (Adelges tsugae) on Eastern Hemlock (Tsuga canadensis) Ecosystems." In Salom, Tigner, and Reardon, *Proceedings*, 42–57.

Everhardt, Gary, Blue Ridge Parkway Superintendent, to Pete Sensabaugh, District Construction Engineer, Virginia Department of Transportation, July 11, 1995. Copy in author's possession.

Federal Register 60, no. 24, February 6, 1995. "Rules and Regulations, Department of the Interior, Fish and Wildlife Service, Final Rule, Endangered and Threatened Wildlife and Plants; Spruce-Fir Moss Spider Determined to Be Endangered."

Feldman, R. S., and E. F. Connor. "Influences of Low Alkalinity and pH on Invertebrate Community Structure: A Controlled Replicated Stream Design." M.S. thesis, Department of Environmental Sciences, University of Virginia, 1985, cited in National Park Service, Air Quality Division and Shenandoah National Park, "Technical Support Document Regarding Adverse Impact Determination for Shenandoah National Park," 22.

Frampton, George T., Jr., Assistant Secretary for Fish and Wildlife and Parks, U.S. Department of the Interior, to Hon. Don Sundquist, Governor, State of Tennessee, March 20, 1996. Copy in author's possession.

General Accounting Office Reports. Resources, Community, and Economic Development Division, Report No. B-257771, October 28, 1994.

Gordon, Christi. "The Cardinal Glass Plant: A Lesson in Civics." *Shenandoah National Park Resource Management Newsletter*, May 1998, 1–2.

Grant, William B., and Orie L. Loucks. "Epidemiological Assessment of Acid Deposition and Ozone Effects on Oak and Hickory Mortality in the Eastern United States." 1997.

Hermann, K. A., ed. *The Southern Appalachian Assessment GIS Data Base CD ROM Set*. Norris, Tenn.: Southern Appalachian Man and the Biosphere Program, 1996.

Hooke, Roger LeB. "Spatial Distribution of Human Geomorphic Activity in the United States: Comparison with Rivers." 1995.

Imhoff, Katherine (executive director, Virginia Commission on Population Growth and Development). Comments in a symposium at Harrisonburg, Va., July 1994.

Irving, Patricia M., ed. *Acidic Deposition: State of Science and Technology*. Vol. 2, *Aquatic Processes and Effects*. National Acid Precipitation Assessment Program. Washington, D.C.: Government Printing Office, 1991.

——, ed. *Acidic Deposition: State of Science and Technology*. Vol. 3, *Terrestrial, Materials, Health, and Visibility Effects*. National Acid Precipitation Assessment Program. Washington, D.C.: Government Printing Office, 1991.

Joseph, David B., and Miguel I. Flores. National Park Service. *Statistical Summary of Ozone Measurements in the National Park System, Quick Look Annual Summary Statistics Reports*. 1994.

Karish, John, Tom Blount, and Bob Krumenaker. Shenandoah National Park, Center for Resources, National Park Service. "Resource Assessment of the June 27 and 28, 1995 Floods and Debris Flows in Shenandoah National Park, Draft of Natural Resource Report." National Parks/CHALSHEN/NRR-96/, January 1997.

Kauffman, John, Larry Mohn, and Paul Bugas. "Effects of Acidification on Bottom Fauna in Saint Mary's River, Augusta County, Virginia." Virginia Department of Game and Inland Fisheries, Staunton, Va.

Keys, James E., Jr., Constance A. Carpenter, Susan L. Hooks, Frank Koenig, W. Henry McNab, Walter E. Russell, and Marie-Louise Smith. *Ecological Units of the Eastern United States, First Approximation*. Department of Agriculture, Forest Service, CD-ROM, 1995.

Knowles, Travis W., Michael A. Steele, and Peter D. Weigl. "Survey of Rare and Endangered Vertebrates of the Blue Ridge Parkway in North Carolina: A Report to the National Park Service." 1989.

Lapin, B. "The Impact of Hemlock Woolly Adelgid on Resources in the Lower Connecticut River Valley." "Report for the Northeast Center for Forest Health Research," Hamden, Conn., cited in Quimby, "Value and Importance of Hemlock Ecosystems," 4.

Lorimer, C. G. "Stand History and Dynamics of a Southern Appalachian Virgin Forest." Ph.D. diss., Department of Forestry and Environmental Studies, Duke University, Durham, N.C., 1976, cited in Meier, Bratton, and Duffy, "Biodiversity in the Herbaceous Layer and Salamanders."

Marland, G., R. J. Andres, and T. A. Boden. "Global, Regional, and National CO_2 Emission Estimates from Fossil Fuel Burning, Cement Production, and Gas Flaring: 1950–1992." Oak Ridge, Tenn.: Oak Ridge National Laboratory Carbon Dioxide Information Analysis Center, 1995, cited in Lester Brown, *State of the World*, New York: W. W. Norton Company, 1996, 29–30.

McNab, W. Henry, and Peter E. Avers, comps. Department of Agriculture, Forest Service. *Ecological Subregions of the United States: Section Descriptions*. 1994.

Miller, H. S., Jr. "The Hemlock Woolly Adelgid, Adelges Tsugae, in Southwest Virginia." Report to the Commonwealth of Virginia, Department of Agriculture and Consumer Services, February 10, 1988, 1–2.

National Acid Precipitation Assessment Program. *1990 Integrated Assessment Report*. Washington, D.C.: Office of the Director, National Acid Precipitation Assessment Program, 1991.

National Center for Health Statistics, 1996 data, cited in Environmental Protection Agency, *Standards for Ozone*, June 1996.

National Park Service. *Handbook 112, Great Smoky Mountains*. Washington, D.C.: Division of Publication, 1981.

———. Letter to the Superintendent, Shenandoah National Park, August 30, 1987.

———. *National Parks for the Twenty-first Century—The Vail Agenda—Report and Recommendations to the Director of the National Park Service*. National Park Service Document #D-726. Post Mills, Vt.: Charles Green Publishing Co., 1994.

———. *Park Science* 16, no. 4 (Fall 1996): 31–32.

National Park Service, Air Quality Division. "Technical Support Document Regarding Adverse Impact Determination for Great Smoky Mountains National Park." 1992.

National Park Service, Air Quality Division and Shenandoah National Park. "Technical Support Document Regarding Adverse Impact Determination for Shenandoah National Park." 1990.

National Park Service, Denver. "Ten-Year Visitation Report with % Change for Annual Data." 1997.

National Park Service, Great Smoky Mountains National Park. "Management Plan," 1995.

National Park Service, Office of Natural Resources. *Natural Resources Assessment and Action Program*. 1988, cited in Wagner et al., *Wildlife Policies*, 45.

National Park Service, U.S. Fish and Wildlife Service. Undated brochure, "Red Wolf—Recovery in the Smokies."

National Research Council. *Science and the Endangered Species Act*. Committee on Scientific Issues in the Endangered Species Act. Washington, D.C.: National Academy Press, 1995.

National Research Council and National Academy of Sciences Committee on Haze in National Parks and Wilderness Areas. *Protecting Visibility in National Parks and Wilderness Areas*. Washington, D.C.: National Academy Press, 1993.

The Nature Conservancy. "Natural Heritage Central Databases (Data on North American Animals Developed in Collaboration with the Association for Biodiversity Information, U.S. and Canadian Natural Heritage Programs and Conservation Data Centres)." 1997.

Nichols, Mary. "Statement by Mary Nichols, EPA Assistant Administrator, Office of Air and Radiation." Environmental Protection Agency, April 2, 1997.

North Carolina Department of Transportation. "The North Carolina Highway Trust Fund and How It Affects You." Brochure. 1991.

———. *U.S. 19 Final Environmental Impact Statement*. #FHWA-NC-EIS-78-09-F. 1983.

———. *Transportation Improvement Plan, 1998–2004.* 1997.

Northeast States for Coordinated Air Use Management, "Air Pollution Impacts of Increased Deregulation in the Electric Power Industry: An Initial Analysis." January 15, 1998.

Noss, Reed, Edward T. Laroe III, and J. Michael Scott. National Biological Service, Department of the Interior. *Technical Report Series,* "Report no. 28, Endangered Ecosystems of the U.S., A Preliminary Assessment of Loss and Degradation." 1995.

Pyle, Charlotte. "Pre-park Disturbance in the Spruce-Fir Forests of Great Smoky Mountains National Park." In White, *Spruce-Fir Ecosystem,* 115–30.

Salom, S. M., T. C. Tigner, and R. C. Reardon, eds. Department of Agriculture, Forest Service, Forest Technology Enterprise Team, Morgantown, West Virginia. *Proceedings of the First Hemlock Woolly Adelgid Review.* 1996.

Saunders, Paul Richard. "Recreational Impacts in the Southern Appalachian Spruce-Fir Ecosystem." In White, *Spruce-Fir Ecosystem,* 100–114.

Senate Documents, 60th Cong., 1st sess., vol. 7, "Report of the Secretary of Agriculture on the Southern Appalachian and White Mountain Watersheds." Doc. no. 91, 1907.

Senate Reports, 60th Cong., 1st sess., vol. 2, "Acquiring National Forests in Southern Appalachian and White Mountains." Report no. 459, April 2, 1908.

Sheffield, R. M., N. D. Cost, W. A. Bechtold, and J. P. McClure. "Pine Growth Reductions in the Southeast." *USDA Forest Service Bulletin SE-83,* Southeastern Forest Experiment Station, Asheville, 1985, cited in Berrang, Meadows, and Hodges, "Overview of Responses," in Fox and Mickler, *Impact of Air Pollutants,* 199.

Sisler, James F., Dale Huffman, Douglas Latimer, principal investigators; William C. Malm, Marc L. Pitchford. *Spatial and Temporal Patterns and the Chemical Composition of the Haze in the United States: An Analysis of Data from the IMPROVE Network, 1988–1991.* 1993.

Soukup, Michael. "Statement by Dr. Michael Soukup, Associate Director, Natural Resource Stewardship and Science." National Park Service, October 28, 1997.

Southern Appalachian Man and the Biosphere. *The Southern Appalachian Assessment.* Department of Agriculture, Forest Service, Southern Region. Vols. 1–5. 1996.

Spitzer, Shane. "New EPA Standards for Ozone." *Shenandoah National Park Resource Management Newsletter,* January 1997, 8.

Stynes, Daniel J. "Visitor Spending and the Local Economy: Great Smoky Mountains National Park." 1992.

Sullivan, Jay, Michael E. Patterson, and Daniel R. Williams. *Shenandoah National Park: Economic Impacts and Visitor Perceptions.* Technical Report National Parks/ MARSHEN/NRTR-93/055. 1992.

Supreme Court of the United States. *Georgia v. Tennessee Copper Company,* 206 U.S. 230; 51 L. Ed. 1038; 27 S. Ct. 618, Argued February 25, 26, 1907, Decided May 13, 1907.

Tennessee Valley Authority, U.S. Army Corps of Engineers, U.S. Fish and Wildlife Service, Final Environmental Impact Statement—Chip Mill Terminals on the Tennessee River. Vol. 1, February 1993.

Thomas Jefferson Planning District Commission, Virginia. *Build-Out Analysis of the Thomas Jefferson Planning District,* 1996.

Thomas Jefferson Sustainability Council, Thomas Jefferson Planning District, Virginia. *Indicators of Sustainability: Interim Report.* 1996.

Thomson, Vivian. *Southern Appalachian Clean Air Partnership.* Department of Agriculture, Forest Service and Department of the Interior, National Park Service. September 1996.

Trijonis, John. "Natural Background Conditions for Visibility/Aerosols." In "Report 24: Visibility: Existing and Historical Conditions, Causes and Effects," ed. John Trijonis, 76. In Irving, *Acidic Deposition*, 76–77.

U.S. Bureau of the Census. "State and County Population 1990 and 1995, Population Distribution and Population Estimates." 1996.

———. "Projections of the Total Populations of States, 1995–2025, series A and B." 1997.

U.S. Congress. 104th Cong., 2d sess., House of Representatives. "Report 104-863—Making appropriations for the Department of Defense for Fiscal Year 1997, Conference Report to Accompany H.R. 3610." 1996.

U.S. Congress, Office of Technology Assessment. *Harmful Non-Indigenous Species in the United States.* OTA-F-565. Washington, D.C.: Government Printing Office, September 1993.

———. *Summary, Harmful Non-Indigenous Species in the United States.* OTA-F-565. Washington, D.C.: Government Printing Office, September 1993.

U.S. Fish and Wildlife Service. "U.S. Fish and Wildlife Service Issues Revised List of 'Candidates' for Endangered Species List." Draft news release, February 1996, 1–4.

U.S. Geological Survey. *Geographic Names Information System.* CD-ROM. February 1995.

U.S. Senate. "Report of the Secretary of Agriculture on the Work of the Biological Survey." Senate Document no. 132, 60th Cong., 1st sess., map following page 14, "Distribution of the Big Wolves," December 21, 1907.

———. "Southern Appalachian and White Mountain National Forests, Report of the Committee on Forest Reservations and the Protection of Game, Supplemental Report to S.4825." 60th Cong., 1st sess., Report 459, Part 2, in Senate Reports, 1907–8, vol. 2, April 2, 1908. Washington, D.C.: Government Printing Office.

———. "Commerce, Science and Transportation Committee Hearing, Jan. 29, 1997." Federal Document Clearing House, Inc., Political Transcripts.

Van Lear, D. H., G. B. Taylor, and W. F. Hansen. "Sedimentation in the Chattooga River Watershed." Department of Forest Resources, Clemson University, Technical Paper No. 19. 1995.

Van Manen, Frank T., Marcus Spicer, and Michael R. Pelton. "Use of Interstate Passageways by Black Bears and Other Wildlife: Final Report to the Nature Conservancy, Tennessee Field Office," Nashville, 1996.

Van Sickle, J., and M. R. Church. "Methods for Estimating the Relative Effects of Sulfur and Nitrogen Deposition on Surface Water Chemistry." U.S. Environmental Research Laboratory, Corvallis, Ore., 1995. Cited in Birnbaum, *Feasibility Study*, 48–51 and app. B.

Watson, J. G., C. F. Rogers, and J. C. Chow. "Pm10 and PM2.5 Variations in Time and Space, DRI Document No. 4204.1F, Report to U.S. Environmental Protection Agency." October 24, 1995, cited in Environmental Protection Agency, *Review of Standards*, 4–13.

Webb, James R., Frank A. Deviney, James N. Galloway, Cheryl A. Rinehart, Patricia

A. Thompson, and Suzanne Wilson. *The Acid-Base Status of Native Brook Trout Streams in the Mountains of Virginia*. Department of Environmental Sciences, University of Virginia, 1994.

White, Peter S. "The Southern Appalachian Spruce-Fir Ecosystem: An Introduction." In White, *Spruce-Fir Ecosystem*, 1–21.

———, ed. *The Southern Appalachian Spruce-Fir Ecosystem: Its Biology and Threats*. National Park Service, Southeast Region, Research/Resources Management Report SER-31, Uplands Field Research Laboratory, Great Smoky Mountains National Park, Gatlinburg, Tenn., 1984.

White, Peter S., and L. A. Renfro. "Vascular Plants of Southern Appalachian Spruce-Fir: Annotated Checklists Arranged by Geography, Habitat, and Growth Form." In White, *Spruce-Fir Ecosystem*, 235–46.

Wigington, P. J., Jr., et al. *Episodic Acidification of Streams in the Northeastern United States: Chemical and Biological Results of the Episodic Response Project*. Office of Research and Development, Environmental Protection Agency, EPA/600/R-93/190, Washington, cited in Birnbaum, *Feasibility Study*, 13.

Yarnell, Susan L. "The Southern Appalachians: A History of the Landscape." Southeastern Center for Forest Economics Research, September 1995 (draft).

Zartman, Charles, and J. Dan Pitillo. "An Inventory of Spray Cliff Plant Communities in the Chattooga Basin." Highlands Biological Station, Highlands, N.C., December 1995.